Theory and Application of Dynamic Wetting of High Voltage Insulator

IEEE Press

445 Hoes Lane

Piscataway, NJ 08854

Theory and Application of Dynamic Wetting of High Voltage Insulator

Bin Cao
Tsinghua University Shenzhen International Graduate School,
Shenzhen, China

Liming Wang
Tsinghua University Shenzhen International Graduate School,
Shenzhen, China

IEEE*Press*
WILEY

Published by John Wiley & Sons, Inc., Hoboken, New Jersey.

Library of Congress Cataloging-in-Publication Data has been applied for:
Print ISBN: 9781394385157
ePDF ISBN: 9781394385171
ePub ISBN: 9781394385164
oBook ISBN: 9781394385188

Cover Design: Wiley
Cover Image: Martin Capek/Shutterstock

Contents

List of Figures

List of Tables

About the Authors

Bin Cao is an assistant professor at the Institute of Advanced Technologies in Energy and Electrical Engineering at Tsinghua University, China. He has published 80 peer-reviewed journal papers and has received multiple prestigious awards, including the 2024 Geneva Invention Exhibition Gold Award and recognition as an Exemplary Individual by the Insulator Standardization Committee in the Power Industry. Dr. Cao is an active member of CIGRE NGN, the Chinese Insulator Standardization Technical Committee, and the IEEE PES High Voltage Surface Discharge Technical Committee, among others.

Liming Wang is a full professor at Tsinghua University, China, and serves as Director of the Institute of Advanced Technologies in Energy and Electrical Engineering. He has authored over 760 peer-reviewed papers, including 104 published in IEEE Transactions journals, and holds 128 patents. Professor Wang serves as Chair of the IEEE PES DC Transmission and Transformation Equipment Technical Subcommittee and the IEEE DEIS Technical Committee on "Discharge in Air at UHV." A frequent speaker at international conferences, he was honored in 2024 with the IEEE DEIS Caixin Sun and Stan Gryzbowski Lifetime Achievement Award.

Preface

High-voltage power transmission lines are critical platforms for global energy interconnection. They play a vital role in optimizing the allocation of renewable energy, improving grid operational efficiency, ensuring energy security, and serve as a key technological pathway for achieving the goals of "carbon peaking" and "carbon neutrality." In high-voltage transmission systems, insulators—being the most widely used key components—are exposed to harsh outdoor environments such as windblown sand, salt fog, rain, and snow.

Traditionally, the design, manufacturing, operation, and maintenance of insulators have relied primarily on **static** parameters such as annual pollution levels, equivalent salt deposit density (ESDD), and pollution flashover voltage. While these indicators help mitigate the risk of widespread pollution flashovers, local flashover events may still occur under increasingly complex and variable climate conditions. According to statistics from the China Electric Power Industry Standardization Committee, hundreds of insulator pollution flashover incidents have occurred domestically in the past five years, resulting in direct economic losses of several billion yuan and posing serious threats to the safe and stable operation of the power grid.

In response to the urgent demands of transmission and substation engineering construction and operation, the Energy and Advanced Electrical Technology team at Tsinghua Shenzhen International Graduate School has adhered to the concept of deep integration between industry, academia, research, and application. Over more than a decade, the team has systematically uncovered the dynamic failure mechanisms of external insulation in power equipment, developed a series of high-performance prevention technologies, and established a new theoretical system and set of standards. These advancements provide strong support for improving the safety and reliability of power equipment's external insulation under extreme operational environments.

This book is based on the latest research progress in this field in recent years. As environmental conditions grow increasingly complex and the demands on external insulation performance become more stringent, significant breakthroughs have been made in studies of pollution wetting characteristics. However, most existing research focuses on static environmental conditions. Systematic studies on wetting mechanisms under dynamic conditions, real-time pollution level monitoring, and correction methodologies remain limited. To fill this theoretical gap, this book systematically reviews and explores the pollution characteristics of high-voltage external insulation under dynamic wetting conditions and their engineering applications, based on the latest research findings. The aim is to provide scientific evidence and practical guidance for the optimal design, operation, and maintenance of external insulation in power systems.

The book not only emphasizes theoretical system development but also closely integrates engineering practice. It thoroughly discusses the application of pollution wetting characteristics in power systems, especially in aspects such as real-time pollution monitoring and correction methods. Detailed descriptions of the latest pollution characterization and correction techniques, monitoring technologies, and representative engineering applications are included, enabling readers to understand the underlying theories and apply the corresponding technical approaches in real-world scenarios.

To enhance readability and practicality, the book is structured with strong systematization and layering, progressing from fundamental to advanced topics. The Introduction discusses the significance of pollution wetting studies on insulators. Chapters 2 to 6 analyze the mechanisms of pollution wetting under saturated, unsaturated, and temperature-difference environments. Chapters 7 to 10 examine wetting and flashover mechanisms on hydrophobic surfaces. Chapters 11 and 14 focus on the application of dynamic wetting theory, including online pollution monitoring and correction methods. Each chapter is relatively self-contained, allowing readers to select sections based on their research or engineering interests, thus improving learning efficiency.

The book's writing was led and primarily authored by Dr. Bin Cao, who was responsible for overall planning, compilation, and the drafting of major chapters. Professors Liming Wang contributed to the development of several chapters and reviewed the technical content throughout the book. We hope this book serves as a valuable reference for researchers, engineers, and university students engaged in the field of high-voltage external insulation, contributing to the reliable operation and technological advancement of China's power system.

Acknowledgments

Flashover caused by surface contamination has long been a classical research topic. Since Obenaus first proposed the pollution flashover model, many pioneering scholars have conducted profound and insightful research in this field. The results presented in this book are built upon the solid foundation laid by these earlier contributions. We acknowledge and deeply respect the invaluable work of all the researchers whose efforts have paved the way for our study.

The authors would like to express their heartfelt gratitude to Prof. Zhicheng Guan, Prof. Zhidong Jia, Assoc. Prof. Ruobing Zhang, Assoc. Prof. Hongwei Mei, Assoc. Prof. Xilin Wang, and Assoc. Prof. Fanghui Yin of the Energy and Advanced Electrical Technology Laboratory at Tsinghua Shenzhen International Graduate School for their generous support throughout this work. We are also deeply thankful to Prof. Zhongdong Wang and Prof. Peter Crossley from the University of Manchester for her valuable guidance. Special thanks are extended to Prof. Boxue Du of Tianjin University for his continuous encouragement during the compilation of this book.

We gratefully acknowledge the assistance provided by Chenlong Zhao, Genggeng Wang, Hongchuan Dong, Kailin Wang, Xu Li, Yidi Shen, Xiaohan Wu, Shichao Shao, Liang Li, and Xindi Zhang during the research process.

This research was supported by the GuangDong Basic and Applied Basic Research Foundation 2026A1515030020, Shenzhen Science and Technology Program JCYJ20250604180237049, National Natural Science Foundation of China 51977118, 52007095, and 52577170, the Shenzhen Outstanding Scientific Innovation Talent Training Program RCBS20200714114956357. We extend our sincere appreciation to all the experts, collaborators, and institutions whose contributions have made this work possible.

Lastly, the authors wish to express their sincere thanks to Mary Hatcher, Christina Spinks, and Nandhini Karuppiah for their dedication, meticulous attention to detail, and tireless efforts in the preparation and finalization of this book.

Acronyms

AC	Alternating Current
API	Application Programming Interface
CFD	Computational Fluid Dynamics
CFLWB	Critical Flashover Length of the Water Band
CIGRE	International Council on Large Electric Systems
DAQ	Data Acquisition
DC	Direct Current
DDGI	Dust Deposit Gauge Index
ESDD	Equivalent Salt Deposit Density
ESSC	Equivalent Soluble Salt Concentration
FFT	Fast Fourier transform
HC	Hydrophobicity Classification
HTV	High Temperature Vulcanized silicone rubber
HV	High Voltage
IC	Ion Chromatography
ICP-AES	Inductively Coupled Plasma Atomic Emission Spectrometry
ICP-OES	Inductively Coupled Plasma Optical Emission Spectroscopy
IEC	International Electrotechnical Commission
IEEE	Institute of Electrical and Electronics Engineers
LCR Meter	Inductance (L), Capacitance (C), and Resistance (R) Meter
MEMS	Microelectromechanical Systems
NSDD	Non-soluble Deposit Density
PID	Proportional-Integral-Derivative
PM10	Particulate Matter $\leq$10 micrometers
PTFE	Polytetrafluoroethylene
RH	Relative Humidity
RTV	Room Temperature Vulcanized
SDD	Salt Deposit Density
SES	Site Equivalent Salinity
THD	Total Harmonic Distortion

1

Challenges in Insulation of Transmission Equipment with Climate Change

The power grid is a critical infrastructure for reliable energy transmission. It underpins economic development, supports daily life, and plays a pivotal role in achieving carbon peaking and carbon neutrality goals. However, in recent years, global climate change and the increasing frequency of extreme weather events have posed unprecedented challenges to energy infrastructure.

Insulators, positioned between the high-voltage components of the grid and ground potential, serve a dual purpose: providing both mechanical support and electrical insulation. As key components directly exposed to outdoor environments for extended periods, insulators form the "first line of defense" for the power grid against climate-induced stress. They are also among the most widely used devices in power systems. By the end of 2023, China had over 7.37 million strings of porcelain insulators, 11.11 million strings of glass insulators, and 14.72 million strings of composite insulators in operation. These insulators collectively form the critical interface between high-voltage transmission lines and the ground. A single pollution flashover event in any one string can cause an unplanned outage of the entire transmission line and may even trigger cascading failures across the grid.

External insulation has long been a key factor influencing the reliability of transmission and transformation equipment. Over the past several decades, researchers worldwide have conducted extensive laboratory and field studies on pollution flashovers. Based on experimental data and operational experience, they have developed design methods to improve the external insulation. However, traditional pollution models and anti-flashover techniques generally treat the site pollution severity (SPS) as a static variable, relying heavily on controlled lab tests and historical field data.

In reality, the contamination and moisture conditions on insulator surfaces are highly dynamic, responding rapidly to environmental changes. In regions with unique environmental conditions or microclimates, surface states of insulators can vary significantly over short periods. This has led to frequent pollution flashover incidents in areas where environmental conditions change abruptly.

In recent years, China has experienced hundreds of such incidents due to a range of factors:

- Sudden transitions during contamination accumulation of insulators (e.g., long dry without rainfall)
- Rapid moisture changes (e.g., persistent "returning dampness" or high humidity)
- Sudden emission of special pollutants (e.g., new chemical plants or firework)
- Simultaneous fast accumulation and wetting (e.g., sea fog)
- Extreme high pollution levels near industrial hubs

These events have resulted in frequent unplanned AC transmission line outages and derated operation of DC lines, causing direct economic losses totaling hundreds of billions of yuan in China and several other countries.

The challenges stem largely from an incomplete understanding of the dynamic mechanisms behind insulator contamination and wetting under complex environmental conditions. There is also a lack of a robust theoretical foundation tailored to natural, variable environments and an absence of truly effective online monitoring and pollution forecasting technologies. As a result, the maintenance of external insulation remains inefficient and imprecise.

According to estimates from relevant authorities, the current global cost of external insulation operation and maintenance is approximately USD 28–34 billion annually. Developing scientific and efficient methods for external insulation operation and maintenance has therefore become one of the most pressing challenges facing the modern power industry (Figure 1.1).

In response to the urgent need for rational design and scientific operation and maintenance of external insulation for transmission and transformation equipment under increasingly complex climatic conditions, this book systematically presents the latest research progress on the dynamic evolution characteristics of surface wetting and pollution-induced discharge of high-voltage transmission and transformation equipment. It explores key technologies for pollution flashover prevention and control under complex operating environments and extreme weather conditions. The aim is to provide valuable references for the

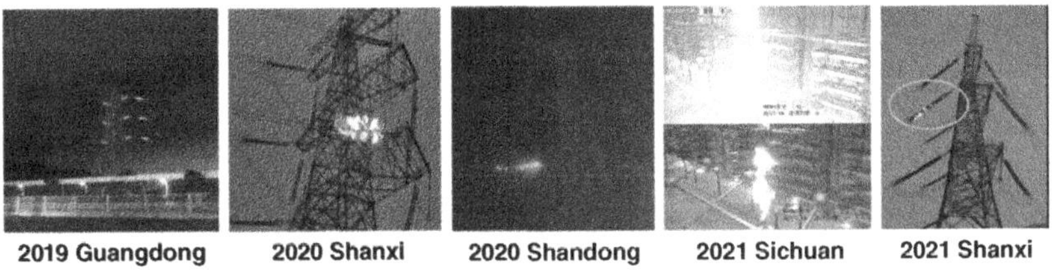

Figure 1.1 Flashover incidents in recent years.

design, selection, monitoring, and maintenance of external insulation systems for high-voltage transmission and transformation equipment in extreme environments, thereby contributing to the safe and stable operation of China's power transmission and transformation infrastructure.

1.1 Classical Theories of Pollution Discharge on Insulators

1.1.1 Pollution Flashover Models

Although pollution flashover incidents were reported as early as the beginning of the 20th century in the United Kingdom, over four decades of research failed to clearly establish the specific physical mechanism behind discharges on wet polluted surfaces. It was not until 1958 that Obenaus first proposed a quantitative analytical model to describe pollution flashover, as illustrated in Figure 1.2. This model suggests that the gradual development of localized arcs along the surface of the pollution layer ultimately leads to the occurrence of a flashover.

After the appearance of a localized arc, the arc forms a series circuit with the remaining polluted layer, which constitutes the classical pollution flashover model. Once discharge occurs, the voltage–current relationship across the insulator can be described by the following equation:

$$U = AxI^{-n} + R(x)I \tag{1.1}$$

where:

U is the applied voltage across the insulator,
A and n are arc characteristic parameters,
x is the arc length (i.e., the length of the local arc),
$R(x)$ is the resistance of the remaining pollution layer,
I is the leakage current.

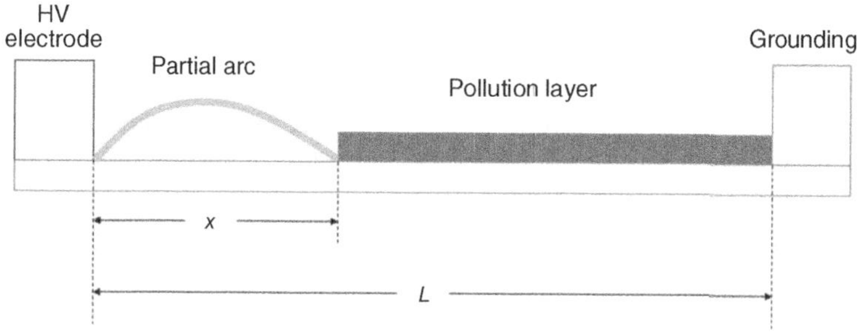

Figure 1.2 Obenaus pollution flashover models.

The first term on the right-hand side of Eq. (1.1), AxI^{-n}, reflects the negative volt-ampere characteristic of the arc. For a given arc length x, the relationship between voltage U and current I forms a "U-shaped" curve, with a minimum voltage needed to sustain the arc, known as the arc sustaining voltage, U_{ext}. If the supply voltage is lower than U_{ext}, the arc cannot be maintained and will extinguish—hence the name extinguishing voltage. Studies have shown that this extinguishing voltage varies with the arc length and reaches a maximum value, known as the critical flashover voltage U_c. This concept forms the basis for much of the experimental and theoretical research into pollution flashover phenomena.

Neumarker was the first to use it to calculate the pollution flashover characteristics of rod-type insulators. However, due to reliance on graphical methods, its application was cumbersome. Alston and Zoledaiowski later developed a more generalized analytical approach through mathematical derivation. Based on the Obenaus model, Alston and colleagues derived expressions for the critical flashover voltage U_c, critical arc length x_c, and critical leakage current I_c, providing a more rigorous theoretical foundation for flashover analysis.

$$U_c = A^{\frac{1}{n+1}} L r_c^{\frac{n}{n+1}} \tag{1.2}$$

$$X_c = \frac{L}{n+1} \tag{1.3}$$

$$I_c = \left(\frac{A}{r_c}\right)^{\frac{1}{n+1}} \tag{1.4}$$

All the above model derivations assume that the conductivity of the residual pollution layer resistance remains constant. However, Wilkins pointed out the variability of the residual pollution layer resistance based on the fundamental model and obtained an analytical expression for the residual pollution layer resistance on the surface of complex-shaped insulators:

$$R(x) = \frac{1}{\pi K} \ln \frac{L-x}{r_0} \tag{1.5}$$

where K is the surface conductivity, and r_0 is the arc root radius.

Furthermore, the critical flashover condition under DC voltage can be derived. Assuming the arc length x is a given fixed value, according to the minimum value theory of the U–I curve, the minimum voltage U_n required to sustain a specific arc length x can be obtained.

$$U_n = \left(1 + \frac{1}{n}\right) (nxA)^{\frac{1}{n+1}} R(x)^{\frac{n}{n+1}} \tag{1.6}$$

U_n is also a function of the arc length x. When the arc length reaches the critical arc length x_c, the voltage required to sustain the arc reaches its maximum value, meaning flashover becomes inevitable. Furthermore, the expression for the critical current can be derived as:

$$I_c = \left[\frac{nx_cA}{R(x_c)}\right]^{\frac{1}{n+1}} \tag{1.7}$$

In the analysis of AC pollution flashover, Claverie and Rizk extended the Obenaus model by proposing that AC flashover not only requires the applied voltage to exceed the minimum extinction voltage but also must satisfy the arc reignition condition. Claverie experimentally summarized the reignition conditions of AC arcs, while Rizk derived the reignition conditions from a mathematical perspective using an energy balance equation.

However, the pollution flashover models developed based on the Obenaus model, whether for DC or AC flashover conditions, have not addressed the development mechanism of surface arcs on the pollution layer. Jolly pointed out that the extinction criterion in Obenaus's model is only a necessary condition for pollution flashover, not a sufficient one. This criterion only indicates that flashover will not occur below the critical voltage U_c, but does not provide the exact flashover voltage. Most studies based on the Obenaus model assume that arc development conditions are automatically satisfied, thus describing only the static process of pollution flashover.

Sundararajan proposed the first dynamic model to describe the evolution of surface arcs on flat plates. However, this model does not consider the unevenness of pollution and moisture on insulator surfaces and assumes the existence of only a single local arc. The dynamic model has the following shortcomings: first, the arc development criterion is not entirely reliable; second, the treatment of the residual pollution layer is overly simplified, assuming uniform and constant conductivity without considering its time-varying nature. Moreover, the residual pollution layer size is assumed to have a linear relationship only with arc length. Rizk also proposed a dynamic model describing arc development. Although this model considered the effects of current, evaporation, and boiling on pollution layer resistance, it did not fully address the arc development conditions and derived the relationship between arc velocity and current mainly based on experimental observations. Experimental results show that surface arc development modes on the pollution layer are diverse, including growth, stagnation, and even retreat.

In summary, existing pollution flashover models are mainly static and neglect arc development conditions. Although some dynamic models have been proposed, they still exhibit significant deficiencies. The fundamental reason is

Table 1.1 Arc characteristic parameter measurement results.

Researcher	*A*	*n*	Pressure p (kPa)	Measurement method
Alston	63	0.76	101.3	AC arc between two metal electrodes in air
B.F. Hampton	530	0.24	101.3	In water vapor conditions
Bergman	84	0.7	70.3	AC arc on flat plate surface
Zhicheng Guan	138	0.69	101.3	DC arc on polluted insulator
Kawamura	63	0.5	101.3	DC arc on triangular glass
Kawamura	35	1.0	13	DC arc on triangular glass

that the arc development mechanism remains unclear. Future research should focus on deeply investigating the arc development mechanism and establishing more comprehensive time-varying models to more accurately describe the arc evolution characteristics. This represents a key research direction in the theory of insulator pollution discharge.

1.1.2 Local Arc Volt-Ampere Characteristics

The study of the volt-ampere (U–I) characteristics of local arcs typically follows the relation: $U = AI^{-n}$. The fundamental problem is to determine the values of the two arc characteristic parameters A and n. However, different researchers have obtained widely varying values for these parameters. A summary of previous research results on arc characteristic parameters is shown in Table 1.1.

The results in the Table 1.1 indicate that the arc characteristic parameters obtained by different experimenters under varying test conditions differ significantly. Therefore, there can be considerable errors when using these arc characteristic parameters for analyzing the local arc volt-ampere characteristics. If one uses local arc characteristic parameters directly for volt-ampere analysis, the results may differ substantially from the actual values.

1.2 Insulator Contamination

The insulator contamination is influenced by a variety of factors. Many researchers have conducted extensive studies on the causes of contamination accumulation. Overall, these influencing factors can be categorized into three main groups: meteorological conditions, atmospheric pollution, and insulator-related parameters.

Meteorological conditions such as wind, precipitation, temperature, and humidity play a significant role. Wind affects the dispersion of pollutants from their sources, the resuspension of dust from the ground, and the airflow characteristics around the insulator surface. Higher wind speeds can transport more contaminant particles toward the insulator in a given time but may also remove some already deposited particles. The direction of the wind may cause uneven contamination distribution on the insulator surface. Rainfall has a strong cleaning effect, significantly reducing surface pollution. Humidity influences not only moisture absorption and flashover risk but also the adhesion of contaminant particles to the insulator surface.

Atmospheric pollution is the primary source of contaminants accumulating on insulator surfaces. Its impact is mainly related to pollutant concentration, composition, and particle size. Soluble salts present in airborne solid or liquid particles can settle on insulators, increasing the surface pollution degree. Particle size affects their behavior in the air and their deposition on the insulator, as different-sized particles are influenced differently by electric fields, wind, and gravity.

Insulator-specific parameters include shape, mounting method, material, and operating voltage. Different insulator shapes create different airflow patterns around them, influencing the probability of particle collisions with the surface. Shape also determines the shading and self-cleaning effects. Different materials exhibit varying adhesion characteristics and responses to rainfall cleaning. Voltage levels affect the strength of the electric field around the insulator, which in turn impacts the behavior of charged particles—especially fine particles that may be influenced more by electrostatic forces. AC and DC voltages also influence contamination differently: under similar conditions, contamination under AC is slightly higher than when de-energized, while DC contamination is significantly higher.

Long-term studies have shown that insulator contamination exhibits strong seasonal patterns. The contamination rate is significantly affected by the time of year. From October to March in North China, reduced rainfall and cold weather, and increased air pollution due to heating by coal lead to a rapid buildup of surface contaminants. This buildup typically peaks in the two to three months before the first significant rainfall in spring. During the rainy season from April to September, heavy rain cleanses the insulator surfaces effectively. In some cases, insulators experience a net reduction in contamination during this period. This seasonal cycle of dry-season accumulation followed by rainy season cleaning results in a long-term saturation effect on insulator contamination.

The contamination characteristics of DC cap-and-pin insulators were investigated over a pollution season. The relationship between equivalent salt deposit density (ESDD) and exposure time followed the equation:

$$\mathrm{ESDD} = Ae^{BM} \tag{1.8}$$

where M is the number of exposure months, and B and A are constants dependent on environmental conditions and insulator structure.

Due to the complexity of contamination mechanisms, it is difficult to isolate the dominant contributing factor. Atmospheric conditions, insulator geometry, electric fields, wind fields, and gravity all interact, making experimental results highly variable and often case-specific. While significant progress has been made in understanding contamination patterns, the fundamental mechanisms remain unclear, posing challenges for improving antipollution flashover strategies.

Currently, there are three main approaches to studying contamination on insulators:

1.2.1 Natural Contamination Stations or Field Measurements

These involve long-term monitoring of insulators under real operating conditions and remain the most widely used method for both AC and DC systems. They provide accurate, practical data but require extended periods to gather sufficient information. Differences in regional climate and geography can also lead to varied conclusions. Since the 1970s and 1980s, numerous natural contamination stations have been established worldwide, offering valuable data for pollution resistance design and maintenance strategies.

1.2.2 Wind Tunnel Testing

Wind tunnel experiments allow researchers to study contamination accumulation under controlled conditions. This method offers a faster way to generate large datasets for systematic analysis. As early as 1979, researchers Mark N. Horenstein and James R. Melcher explored the impact of electric fields on particle deposition in DC high-voltage environments. Their findings revealed that particles smaller than 10 μm are significantly influenced by electrostatic forces, whereas larger particles are dominated by aerodynamic drag. Wind tunnel results show that the contamination rate is closely related to wind speed and is typically higher on the top surface of the insulator. The structural design also matters: initially clean insulators accumulate contaminants slowly, but the rate increases after a thin pollution layer forms. Composite insulators typically perform better than ceramic ones in these tests. Additionally, larger umbrella sheds accumulate less contamination than smaller ones, while dust-laden airflow speed and exposure duration are key factors.

1.2.3 Computational Fluid Dynamics

Advances in computer technology have made computational fluid dynamics (CFD) a valuable tool for simulating airflow and particle movement around insulators. Using numerical models like random trajectory or random walk models, researchers can track particles in turbulent flow and predict how they interact with the insulator surface. By calculating collision rates, SPS of insulator can be estimated under various conditions. CFD offers high repeatability and control, enabling analysis of the effect of individual variables. However, its accuracy depends heavily on the precision of the mathematical models and parameter settings, making experimental validation essential.

1.3 Insulator Wetting

The wetting, along with the insulator pollution accumulation, is both critical stages in the process of pollution flashover. Flashover will not occur in the absence of either condition, making both equally important from the perspectives of flashover mechanisms and antipollution strategies. Factors influencing the moistening characteristics of insulators fall into three main categories: environmental conditions (fog, rain, dew, icing, etc. and their related parameters), pollution-related factors (types and contents of soluble salts and inert substances), and the insulator's own characteristics (type, suspension method, and surface temperature). The amount of water that saturates the insulator surface is significantly affected by the mechanism of pollution layer moistening. Numerous studies have explored the moisture characteristics of insulators through parameters such as leakage current and pollution flashover voltage, achieving substantial progress. In the early 1980s, Canadian researcher M. Leclerc conducted foundational studies on surface moistening mechanisms, classifying the modes into droplet impact, condensation, adsorption, and molecular diffusion. Building on this framework, subsequent studies further examined how insulator pollution degree, ambient temperature, and other factors influence surface wetting.

1.3.1 Droplet Impact

Under conditions of heavy fog or light rain, when relative humidity approaches 100%, small droplets collide with the insulator surface due to gravity, wind, and electric fields, leading to rapid wetting of the surface pollution layer. The rate of this process is mainly determined by droplet density, velocity, and trajectory. Different types of fog, therefore, significantly influence impact moistening behavior. Artificial fogs used in pollution tests include steam fog, cold fog, and hybrid fog. Cold fog is typically produced via ultrasonic atomization or aspiration, while steam fog is generated by heating deionized or tap water in boilers. Hybrid fog combines both. Large-scale artificial pollution test chambers in China and abroad generally use steam or hybrid fog.

1.3.2 Molecular Diffusion

Moistening by molecular diffusion involves the transformation of airborne water vapor into liquid phase on the pollution layer. This process is often analyzed using the two-film theory to quantify mass transfer flux and rate. Since the vapor pressure of deionized water is higher than that of the pollutant solution, water molecules in the atmosphere continuously diffuse into the pollution layer. The mass transfer flux J_A of water molecules through the gas and liquid films is given by:

$$J_A = k_L \alpha (C_A^* - C_A) \tag{1.9}$$

where α is the surface area per unit mass, k_L is the film mass transfer coefficient, C_A^* is the equilibrium concentration of component A, and C_A is the actual concentration.

1.3.3 Adsorption Moistening

Naturally accumulated pollutants on insulator surfaces typically consist of a mix of soluble and insoluble substances. Some materials—particularly soluble salts, sugars, and compounds like caprolactam—exhibit strong hygroscopicity. As a result, even at relatively low humidity, the pollution layer can absorb water molecules from the air. Several surface flashover events in China have been linked to adsorption moistening. For instance, shortly after the commissioning of the 500 kV Gangcheng Substation in Zhanjiang in 2006, severe surface discharges occurred despite humidity being below 80%. Subsequent analysis revealed high concentrations of strongly hygroscopic substances like sugars and aluminum phosphate in the surface contaminants.

1.3.4 Condensation

When the surface temperature of an insulator drops below the dew point for the current temperature and humidity conditions, rapid condensation of water vapor can occur, leading to fast moistening of the pollution layer. The dew point temperature D_p is calculated as:

$$H = [\lg(RH) - 2]/0.4343 + 17.62T/(243.12 + T) \tag{1.10}$$

$$D_p = 243.12H/(17.62 - H) \tag{1.11}$$

where RH is the relative humidity and T is the ambient temperature.

In desert regions, early-morning flashovers may be attributed to thermal gradient moistening. Although the air humidity is usually low, the insulator surface temperature can drop below the ambient temperature at night, triggering condensation. Experimental studies have explored the relationship between temperature gradients and peak leakage current, but data on actual moisture accumulation during this process remains scarce.

1.4 Characterization Parameters for External Insulation

Selecting appropriate parameters to characterize the external insulation condition of transmission lines is essential for accurately assessing their operating status and preventing pollution flashover accidents. The external insulation performance of an insulator is primarily determined by the degree of surface contamination and moisture absorption. At present, the main methods for characterizing insulator surface contamination include ESDD, surface conductivity, leakage current, pollution flashover voltage, and flashover voltage gradient. These parameters can generally be divided into two categories: measurements of contamination directly on the insulator surface, and assessments based on environmental pollution monitoring.

Environmental monitoring focuses on tracking meteorological and atmospheric parameters in the insulator's surroundings to indirectly infer surface pollution levels. Common techniques include the use of directional dust deposit collectors and analysis of airborne pollution samples. However, because this approach only provides indirect information about the actual condition of the insulator surface, it has seen limited application in domestic power systems.

Parameters used to characterize the insulator contamination include ESDD, pollution layer conductivity, localized surface conductivity, leakage current, dust deposit density, and flashover voltage gradient. In addition to these, some research has introduced further indicators such as the volume conductivity of collected pollution samples, the characteristics of leakage current pulses, and the frequency of flashover events. These extended parameters offer a more comprehensive understanding of external insulation behavior under both artificial and natural contamination conditions.

1.4.1 ESDD

ESDD is a key parameter for evaluating the contamination level of insulators. It represents the amount of NaCl-equivalent conductive salt deposited per unit surface area of the insulator. The non-soluble portion of the contamination can be measured and calculated as the Non-Soluble Salt Density (NSDD).

Under artificial pollution test conditions, the flashover voltage of an insulator typically has a negative power-law relationship with its surface ESDD. However, factors such as the magnitude of the NSDD, the type of salt, and the distribution of the contamination layer can all influence the relationship between ESDD and flashover voltage. To account for the influence of different salt types and dust components on the insulator's flashover voltage, a multiparameter correction is introduced based on Eq. (1.12), resulting in the following form:

$$U = U_0 \times \prod_i K_i \qquad U_0 = A \times \mathrm{ESDD}^{-\beta} \tag{1.12}$$

where:

U is the flashover voltage of the contaminated insulator;
U_0 is the 50% flashover voltage under ideal conditions (uniform pollution on both upper and lower surfaces, with NaCl as the sole conductive material in the contamination);
k_i are correction factors for various influencing parameters;
A is a discharge constant determined by the insulator's material and structure;
β is a weighted average exponent related to the surface conductivity.

Through experimental studies, domestic and international research institutions have obtained empirical correction factors (k_i) under various conditions, including different levels of soluble salt density, dust content, pollution composition, and pollution distribution. In practice, insulators with the same ESDD under natural and artificial pollution often exhibit different flashover voltages. Even after applying corrections, a completely equivalent relationship between the two is not achieved. Furthermore, when using ESDD to evaluate contamination levels, if the composition of natural pollution or the distribution of pollutants on the insulator surface is unknown, it remains difficult to accurately estimate the flashover voltage. This highlights a limitation of ESDD: while useful, it is not an ideal or comprehensive parameter for characterizing insulator contamination.

1.4.2 Insulator Flashover Voltage and Flashover Voltage Gradient

Flashover voltage is the most direct physical quantity used to characterize the external insulation performance of an insulator. One of the main objectives of contamination level assessment is to estimate the insulator's flashover voltage, thereby providing a basis for operational decision-making. Research has shown that under typical conditions, there exists a correlation between the flashover voltage gradient E (kV/m) and a pollution severity index H, as expressed in the following relationship.

$$E \propto H^{-p} \tag{1.13}$$

From a mathematical perspective, the parameter p is regarded as a weighted average representing the electrolyte surface characteristics. Researchers have provided values for these parameters under different types of insulators and test conditions. Experimental results demonstrate that this relationship generally holds for various types of artificial pollution tests. However, it has also been observed that the regression coefficients obtained from natural pollution tests often differ from those derived from artificial pollution tests. Thus, the mathematical equivalence between natural and artificial contamination conditions remains unresolved, presenting an ongoing challenge in pollution flashover research.

1.4.3 Leakage Current

Extensive research has been conducted both domestically and internationally on the relationship between leakage current and the external insulation condition of insulators. Studies indicate that leakage current is influenced by both environmental factors (such as relative humidity and temperature) and surface contamination characteristics (such as contaminant composition, degree of pollution, and uniformity). Due to its strong correlation with external insulation performance and its capability for continuous online monitoring, leakage current has become widely applied in insulation diagnostics.

Parameters such as the leakage current value immediately prior to flashover, the number of leakage current pulses, and the maximum leakage current under operating voltage can be used to characterize the insulation condition. Among these, the maximum leakage current under operating conditions is currently the most commonly used indicator. This is because:

- The number of leakage current pulses has limited correlation with flashover voltage.
- The pre-flashover leakage current is difficult to measure in real time.

Current leakage current prediction methods can be broadly classified into heuristic (technical) methods and quantitative modeling methods:

- *Heuristic methods*, such as expert judgment, rely heavily on the experience and technical skill of the evaluator.
- *Quantitative methods* build predictive models, such as multiple regression or artificial neural networks, to forecast surface leakage current.

Among these, neural network models are the most widely used. However, most of these models are based on laboratory data, and their input variables are relatively limited. As a result, they often fail to account for the complexity of real-world operational environments, which limits their practical application in the field.

1.4.4 Surface Conductivity

The surface conductance of an insulator is defined as the ratio of the power-frequency current flowing along the surface of the insulator to the voltage applied across its terminals. Surface conductance reflects the overall insulation condition of the insulator surface, accounting not only for the contribution of conductive components within the contamination but also for the moisture absorption effects of the non-soluble components.

Since the conductance of the pollution layer is influenced by the shape and dimensions of the insulator, a normalized parameter called surface conductivity (K) is introduced to facilitate comparisons between different types of insulators. It is defined as:

$$K = f \times G = f\frac{I}{U} \tag{1.14}$$

where:

U is the voltage applied across the insulator string,
I is the leakage current flowing along the surface,
f is a shape factor that reflects the geometric configuration of the insulator.

The shape factor f is given by:

$$f = \int_0^L \frac{dx}{\pi D(x)} \tag{1.15}$$

where:

L is the creepage distance of the insulator,
$D(x)$ is the diameter at a point x along the creepage path.

To accurately measure the overall surface conductivity, an appropriate voltage must be applied across the insulator string. This voltage should be high enough to reflect the insulation condition under operating conditions, yet the duration of the voltage application must be short to minimize the effects of heating and partial discharge. Currently, there is no internationally standardized procedure regarding the voltage magnitude or duration for surface conductivity measurements. Table 1.2 summarizes the voltage application parameters used by different research institutions for measuring overall surface conductivity.

Due to the complexity of instruments and methods required to measure the overall surface conductivity of an insulator string, as well as the associated costs, such measurements are not economically viable for large-scale deployment.

To address this limitation, the concept of local surface conductivity was introduced. From a physical standpoint, overall and local surface conductivity

Table 1.2 Voltage requirements for surface conductivity measurement by different research institutions.

Standard/institution	Requirements
German standard	24 kV per meter of discharge distance
IEC 60507	At least 700 V per meter of creepage distance (as specified for artificial pollution tests)
CIGRE	30 kV per meter of insulator length
Standards from former Soviet Union	Measurement conducted at 80% of the flashover voltage

represent the same underlying parameter—surface conductivity—but differ primarily in measurement approach. Measuring overall surface conductivity typically requires applying voltage across the entire insulator string, which involves full-string wetting and high-voltage equipment. In contrast, local surface conductivity measurements involve wetting only a localized area of the insulator surface, then measuring the conductivity in that region to estimate the local surface conductivity.

This localized method offers several advantages:

- It reflects the degree of local contamination,
- It enables analysis of pollution distribution patterns,
- It allows researchers to observe contamination accumulation and loss trends without damaging or altering the pollution layer.

Because of these benefits, local surface conductivity has been widely adopted under laboratory conditions. To estimate the overall insulation performance using partial measurements, researchers commonly use direct averaging or weighted averaging. However, under conditions of uneven pollution distribution or unclear distribution patterns, these averaging methods can introduce errors when evaluating the overall performance of the insulator string. Currently, there is no well-developed or standardized equipment for surface conductivity measurement.

1.5 Pollution Monitoring Technology

The current selection process for external insulation on transmission lines begins with plotting pollution maps based on ESDD measurements. Insulator configurations are then determined based on specific creepage distance and insulator design parameters. However, with the development of industry and agriculture, the environmental conditions surrounding transmission lines have changed significantly, while pollution maps are not updated in real time. Moreover, the accumulation of surface pollution on insulators requires a certain contamination period, meaning that ESDD measurements cannot reflect the insulation status of insulators in real time. As a result, flashover incidents caused by insufficient creepage distance still occur. Implementing online monitoring and assessment of insulator pollution conditions can guide daily maintenance of transmission lines, reduce the randomness in the application of antipollution measures, improve productivity, and lower labor intensity. Therefore, online monitoring of pollution status is an inevitable trend in the development of modern power grids.

Multiple technologies are available for online monitoring of external insulation, including leakage current monitoring, partial discharge detection, infrared thermography, ultraviolet imaging, and comprehensive assessments based on meteorological and environmental parameters. Among these, leakage current monitoring is the most commonly used method. Leakage current monitoring offers the advantage of continuous online observation. The real-time measurements correlate with the amount of surface contamination and moisture content on the insulator, allowing for a degree of assessment of the pollution status. However, the development of pollution assessment systems based on leakage current faces challenges. One major issue is the short lifespan of sensors, which cannot operate reliably over long periods. Another is that leakage current is influenced by multiple variables—including line voltage, pollution levels, and moisture content of the pollution layer. Existing evaluation models—such as neural networks, wavelet transform models, time-domain and frequency-domain analysis models, phase angle models, and fundamental wave models—mainly analyze the trend of leakage current without establishing a direct correlation with the flashover voltage of insulators.

Therefore, it can be concluded that previous monitoring systems primarily served the purpose of observing data, without linking surface pollution levels to predefined safety thresholds. They could not evaluate pollution status and flashover risk warnings. More accurate monitoring systems and more intelligent expert evaluation platforms are urgently needed. A key step in improving online pollution monitoring systems is resolving the relationship between real-time measurements and the worst-case scenario values, as flashover typically occurs only when the insulator surface is fully saturated with moisture, not under normal operational (non-saturated) conditions.

Beyond leakage current monitoring, other pollution detection techniques have also been explored. For example, infrared thermography is based on the fact that leakage current on wet, polluted insulators generates Joule heating, which increases surface temperature. The more severe the contamination, the more significant the temperature rise. However, this heating phenomenon is heavily influenced by ambient temperature, humidity, and operating voltage. Studies have shown that clean insulators and dry, polluted insulators do not exhibit noticeable temperature changes during operation, while wet, polluted insulators show significant variations.

To enhance infrared-based detection, researchers have combined image processing techniques with infrared imaging to extract relevant features. For instance, a wavelet-based generalized Gaussian distribution and maximum a posteriori estimation method has been proposed for infrared image denoising, which outperforms traditional methods. For infrared images where insulator-background contrast is weak, some researchers have divided the image into background and insulator regions, used grayscale histogram metrics to compute intra-class absolute differences, and applied pulse-coupled neural networks with iterative processing to achieve optimal image segmentation. Additionally, combining hyperspectral imaging and infrared thermography has proven effective. Hyperspectral data provide information on surface contamination composition and concentration, while infrared imaging reveals thermal characteristics. Together, these parameters can be used to establish models that classify pollution levels, expanding the scope of pollution state detection.

Apart from online monitoring, offline computational methods also play a critical role in pollution assessment. These methods correlate environmental monitoring data, such as the concentrations of fine particulate matter with aerodynamic diameters of 2.5 μm or less (PM2.5) with measured ESDD to build pollution prediction models. It helps devise anti-flashover strategies for regional grids, thereby improving system reliability.

In summary, both domestic and international efforts have made significant progress in insulator contamination monitoring, resulting in the development of a variety of technical solutions. However, each method has its limitations. Future research should focus on enhancing the accuracy, reliability, and adaptability of monitoring technologies to better ensure the safe operation of power systems.

1.6 Main Content of the Book

As climate conditions grow increasingly complex, there is a pressing need for the rational design, scientific operation, and effective maintenance of external insulation in high-voltage transmission and transformation equipment. In response, this book systematically presents the latest research on the dynamic evolution of surface wetting and pollution-induced discharge phenomena affecting external insulation systems. We hope it could offer a comprehensive analysis of the challenges that climate change poses to insulation performance. It explores the intricate relationships between environmental factors, material properties, and surface contamination, all of which have a direct impact on insulation reliability and service life.

The content is divided into four major thematic sections, as shown in Figure 1.3.

1. This section (Chapters 2–6) delves into how various environmental factors—such as saturated and unsaturated humidity, temperature variations, fog, and salt dissolution—impact the wetting behavior of contaminated insulation surfaces. These mechanisms are crucial to understanding the formation and persistence of conductive paths that lead to insulation failure.
2. Covered in Chapters 7–10, this section focuses on the hydrophobic properties of insulation materials and their effects on contamination wetting. It investigates how different levels of surface hydrophobicity and material characteristics influence flashover behavior and how these effects are modulated by other factors.
3. Chapters 11–12 introduce modern techniques for monitoring contamination on insulation surfaces. Special attention is given to real-time monitoring systems that operate under various humidity conditions, offering practical insights into predictive maintenance and reliability improvements.
4. The final technical section (Chapters 13–14) presents advanced analysis and correction strategies for soluble salt contamination on insulation surfaces. It includes spatial and temporal analysis methods and solutions to improve contamination assessments under changing environmental conditions.

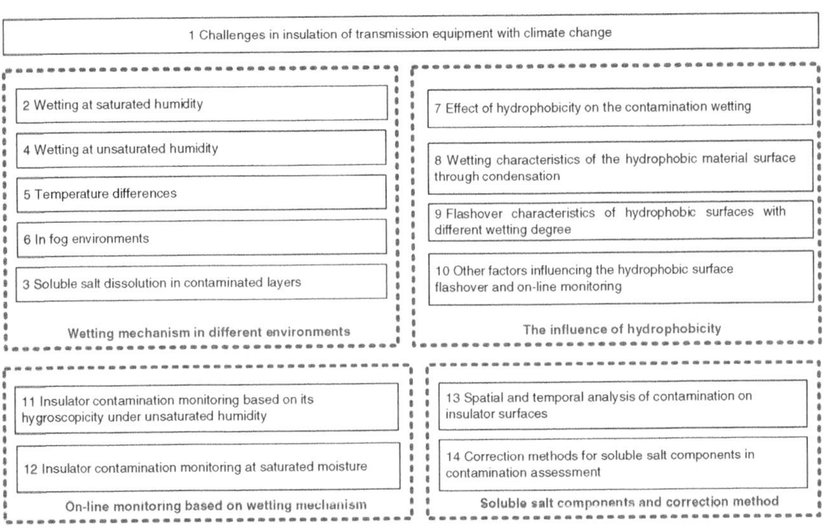

Figure 1.3 The book's graphical summary.

Bibliography

Bojovschi, A., Rowe, W., and Wong, K. The influence of hanging water droplets on discharge activity, application to high voltage insulators, Appl. Phys. Lett., 2011, 98 (9), 091504.

Bruggeman, P., Guns, P., Degroote, J. et al. Influence of the water surface on the glow-to-spark transition in a metal-pin-to-water electrode system, Plasma Sources Sci. Technol., 2008, 17 (4), 045014.

Fujii, O., Honsali, K., Mizuno, Y. et al. A basic study on the effect of voltage stress on a water droplet on a silicone rubber surface, IEEE Trans. Dielectr. Electr. Insul., 2009, 16 (1), 116–122.

Fujii, O., Honsali, K., Mizuno, Y. et al. Vibration of a water droplet on a polymeric insulating material subjected to AC voltage stress, IEEE Trans. Dielectr. Electr. Insul., 2010, 17 (2), 566–571.

Guan, Z., Yu, S., and Jia, Z. The pollution and hydrophobicity properties of RTV coating, High Volt., 1993, 19 (4), 28–32.

He, L. and Gorur, R.S. Source strength impact analysis on polymer insulator flashover under contaminated conditions and a comparison with porcelain, IEEE Trans. Dielectr. Electr. Insul., 2016, 23 (4), 2189–2195.

Jiang, X.L., Zhao, S.H., Xie, Y.B. et al. Study on fog flashover performance and fog-water conductivity correction coefficient for polluted insulators, IET Gener. Transm. Dis., 2013, 7 (2), 145–153.

Liang, X. and Li, S. Looking to the future of composite insulators, 2015 INMR World Congress, Munich, Germany, 2015, pp. 18–21.

Liang, X., Gao, Y., Wang, J. et al. Rapid development of silicone rubber composite insulator in China, High Volt., 2016, 42 (9), 2888–2896.

Liu, Y. and Du, B.X. Energy eigenvector analysis of surface discharges for evaluating the performance of polymer insulator in presence of water droplets, IEEE Trans. Dielectr. Electr. Insul., 2014, 21 (6), 2438–2447.

Liu, Y., Du, B., and Farzaneh, M. Self-normalizing multivariate analysis of polymer insulator leakage current under severe fog conditions, IEEE Trans. Power Deliv., 2017, 32 (3), 1279–1286.

Liu, Y., Wu, Y., and Du, B.X., Dynamic formation mechanism of water droplet and induced surface discharges on SiR composites, High Volt., 2019, 4 (1), 59–64.

Nazemi, M.H. and Hinrichsen, V. Experimental investigations on water droplet oscillation and partial discharge inception voltage on polymeric insulating surfaces under the influence of AC electric field stress, IEEE Trans. Dielectr. Electr. Insul., 2013, 20 (2), 443–453.

Obenaus, F. Contamination flashover and creepage path length, Deutsche Elektrotechnik, 1958, 12, 135–136.

Salem, A.A., Rahman, R.A., Kamarudin, M.S. et al. Factors and models of pollution flashover on high voltage outdoor insulators: review, IEEE Conference on Energy Conversion (CENCON), 2017, pp. 241–246.

Soares, D., Mendonça, S., Neto, E. et al. Electrical field on non-ceramic insulators and its relation to contact angles for constant volume droplets, J. Electrost., 2016, 84, 97–105.

Wilkins, R. and Al-Baghdadi, A.A.J. Arc propagation along an electrolyte surface, Proc. Inst. Electr. Eng., 1971, 118 (12), 1886–1892.

Wilkins, R. Flashover voltage of high-voltage insulators with uniform surface-pollution films, Proc. Inst. Electr. Eng., 1969, 116 (3), 457–465.

Yang, L. Evaluation of wetting condition and its effects on pollution flashover voltage of aerodynamic insulators, IEEE Trans. Dielectr. Electr. Insul., 2016, 23 (5), 2875–2882.

Zhang, C.Y., Hu, J.Y., Li, J.C. et al. Experimental study on the contamination deposition characteristics of insulators in a fog–haze environment, IET Gener. Transm. Dis., 2018, 12 (2), 406–413.

Zhang, C.Y., Wang, L.M., Guan, Z.C. et al. Pollution flashover performance of full-scale ±800 kV converter station post insulators at high altitude area, IEEE Trans. Dielectr. Electr. Insul., 2013, 20 (3), 717–726.

2

Wetting Behavior of Insulator Contamination at Saturated Humidity

2.1 Introduction

Analysis of the wetting characteristics of insulator surfaces is essential for understanding the mechanisms of pollution flashover. Chapters 2–6 focus on the wetting mechanisms of contaminated insulation surfaces under different environmental conditions. Firstly, the fundamental evolution of surface wetting behavior is examined by analyzing the saturated water absorption characteristics of the contamination layer during the wetting process. Subsequently, the dissolution behavior of soluble salts in the contamination layer is systematically analyzed under different moisture contents, revealing its influence on surface wetting characteristics. Finally, the impacts of temperature variation and fog environments on the wetting dynamics of contaminated surfaces are discussed, and a quantitative description method for fog concentration applicable to external insulation equipment is proposed. Together, these chapters provide a systematic understanding of the environmental wetting mechanisms that govern the initiation and development of pollution flashover.

First, the chemical composition of the contamination on the naturally polluted insulator is determined by its surrounding pollution sources. The soluble part in the contamination contains not only some soluble salts with large solubility, but also a certain amount of slightly soluble salts such as $CaSO_4$. In IEC 60815-1, the equivalent salt deposit density (ESDD) is used to characterize the pollution degree of insulators. However, due to the limited water content on the insulator surface under saturated moisture condition, some of the slightly soluble salts, which have a secondary influence on the pollution flashover voltage, are incompletely dissolved. Therefore, the analysis of water absorbed by the contamination layer is also very meaningful to study the contribution of slightly soluble salts to pollution flashover voltage.

The wetting characteristics of polluted insulators have been studied by many researchers. Leclerc categorized the wetting processes of a contaminated insulator surface into several classes, namely, collision of water droplets with the insulator surface, condensation, the hygroscopic behavior of the surface deposit, and molecular diffusion. However, the saturated water absorption of the contamination layer was not given in his study. Some researchers measured the saturated water attached to the insulator surface after wetting. Ling concluded that saturated water content of the XP-70 insulator was about 10 mL after wetting using an injection to spray water on the polluted insulator surface. Huang found that, in the steam fog, there was a linear relationship between saturated water absorption of the insulators and Non-soluble Deposit Density (NSDD). However, as the saturated water absorption of the contamination layer is very little for insulators, the method of weighing measurement might be inaccurate.

In this chapter, the saturated water absorption of contamination layer on the specimen surface was measured and analyzed. For a given sodium chloride (NaCl) solution, the ratio of conductive particles to water molecules is constant in the salt fog generated. Therefore, the surface conductivity of the contamination layer was linear with the deposition mass of salt fog particles under saturated moisture condition. In this book, a new method was proposed to analyze the saturated water absorption of the contamination layer based on surface conductivity. By this method, the influence of NSDD and the inclination angle of the specimen on the saturated water absorption of the contamination layer were analyzed.

2.2 Example Arrangement

2.2.1 Test Specimens, Facilities & Procedure

For contamination layer that is deposited on the insulator surface, the water content it absorbs is very little. Therefore, it might be difficult to get an accurate measurement of the saturated water absorption by weighing the mass of the insulator before and after wetting. For example, for a typical XP-70 insulator whose weight is about 4.6 kg, the measured values of saturated water content absorbed by the contamination layer are widely scattered, ranging from 8 to 20 g. Second, the wetting status is different at different locations, especially for the contamination layers on the bottom surface of the insulator compared with that on the top surface. Finally, to keep the power system running, it is unrealistic to measure the saturated water content by removing the insulator from the transmission line tower.

To reduce contamination deposition and facilitate rain washing, the insulator usually has a curved shape, as shown in Figure 2.1. This chapter preliminarily

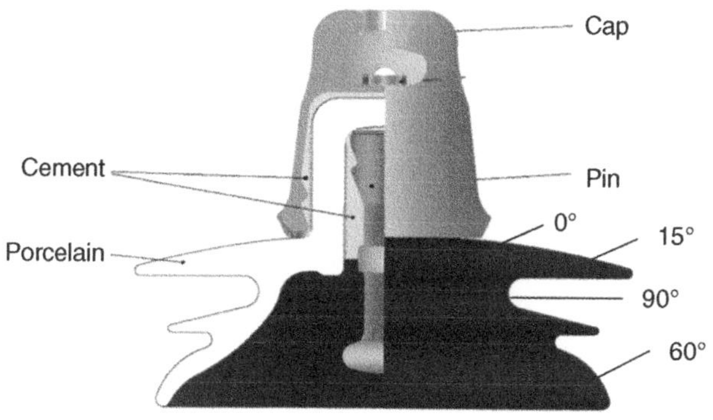

Figure 2.1 The schematic diagram of a typical insulator.

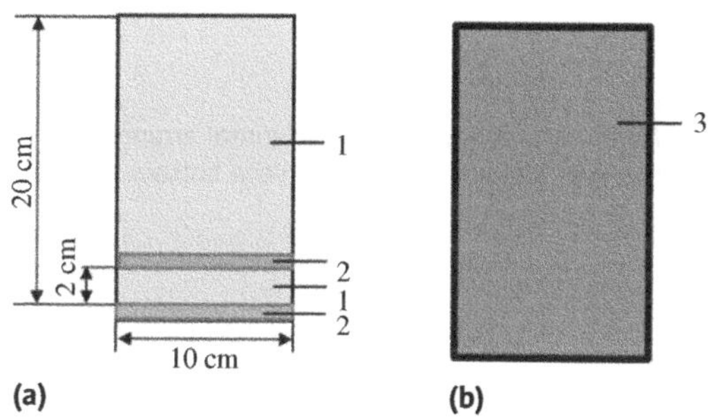

Figure 2.2 Schematic diagram of the glass plate specimen: (a) contaminated surface of the glass plate specimen; (b) the other surface of the specimen (coated with RTV). 1: Contamination layer; 2: parallel conductive tapes; 3: RTV coating. *Source*: Cao et al. (2019)/with permission of IEEE.

studied the saturated water absorption at different locations on the insulator surface, which can provide reference for insulator pollution flashover modeling.

The surface inclination angle at the insulator surface changes continuously. In this chapter, a method to analyze the water absorption by the contamination layer during the wetting process was proposed. To simplify the measurement, contaminated glass plate specimens placed at various inclination angles were used to measure the saturated water absorption at different inclination angles. The schematic diagram of the glass plate specimen is illustrated in Figure 2.2. One surface of the specimen was contaminated, and the other surface was coated with RTV. The specimen has a length of 20.5 cm and a width of 10 cm. A pair of parallel conductive tapes with dimensions of 10 cm (length) × 0.5 cm (wide) × 0.01 cm (thickness) were struck to the contaminated specimen surface.

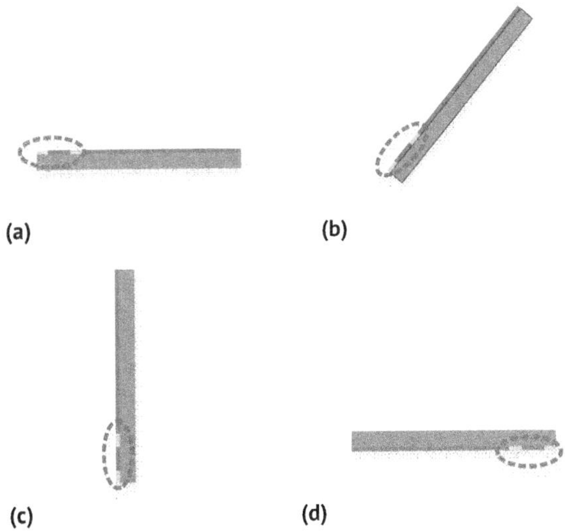

Figure 2.3 The glass specimens in different arrangements: (a) horizontal arrangement; (b) inclined arrangement of top surface; (c) vertical arrangement; (d) the bottom surface in horizontal arrangement.

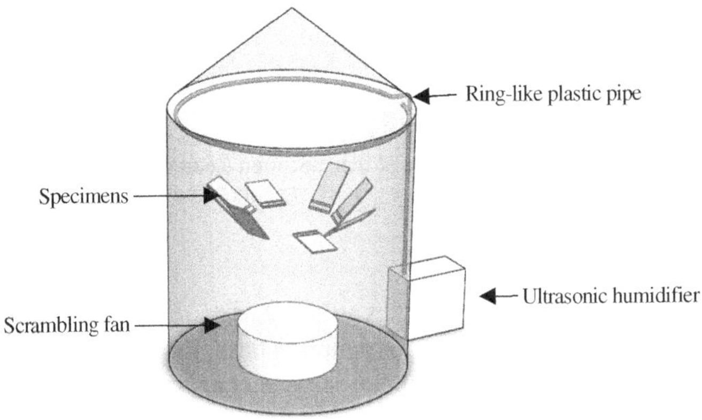

Figure 2.4 Structure diagram of the testing system.

In this chapter, specimen arrangements including horizontal arrangement, inclined arrangement, and vertical arrangement were investigated, as shown in Figure 2.3. The measured area of the contaminated surface of the specimen is circled. When placed inclined as shown in Figure 2.3b, it will take less time for the measured area to get the saturated state.

To wet the specimen, the tests were carried out in the salt fog chamber. The testing system is illustrated in Figure 2.4. The salt fog chamber had a base radius of

Figure 2.5 The appearance of contaminated specimen.

100 cm and a height of 120 cm, and the effective test space was 25–95 cm height above the chamber floor. In order to avoid droplets that may be formed on the top of the chamber falls on the surface of the test specimens, the chamber top was designed as a cone. The salt fog was generated from NaCl solution by an ultrasonic humidifier. After the generation, the salt fog particles were channeled into the fog chamber though a pipe. A ring-like plastic pipe full of small holes was installed in the fog chamber at a height of 95 cm and was responsible for spreading salt fog into the chamber. A scrambling fan was installed at the center of the fog chamber floor to make ensure that the distribution of salt fog was as uniform as possible. The specimens were placed on a supporter made by 2 mm plastic rods as illustrated in Figure 2.5.

In the tests, an UNI-T UT612 LCR meter, which is a top tool for measuring inductance, capacitance and resistance with minimum fuss and hassle, was used to measure the surface resistance (inverse of surface conductivity) of the wetted contaminated layer between two parallel conductive tapes as shown in Figure 2.6. The parameters of the LCR meter are listed in Table 2.1. In the tests, the test frequency of LCR meter was set at 100 Hz. As the voltage applied to the specimens was small and the resistance of the contamination layer was large, the thermal effect caused by the leakage current can be safely ignored.

As shown in Figure 2.7, a certain amount of salt fog with given conductivity was channeled into the fog chamber first. Meanwhile, the surface conductivity of the specimen was measured by UNI-T UT612 LCR meter. According to the relationship between water absorption and the measured surface conductivity of the

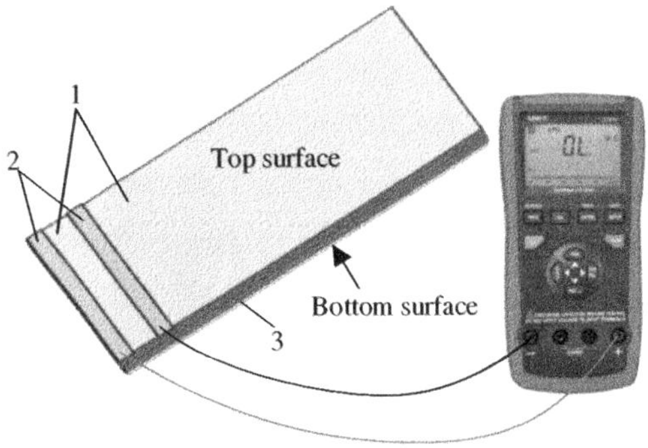

Figure 2.6 Measurement of the resistance by LCR meter (not to scale). 1: Contamination layer; 2: parallel conductive tapes; 3: RTV coating. *Source*: Cao et al. (2019)/with permission of IEEE.

Table 2.1 Parameters of the UNI-T UT612 digital LCR meter.

Items	Parameter
Range	20 Ω ~ 200 MΩ
Accuracy	±(0.5% Range + 5 Ω)
Operation voltage	0.6 V_{rms}
Test frequency	100 Hz ~ 10 kHz

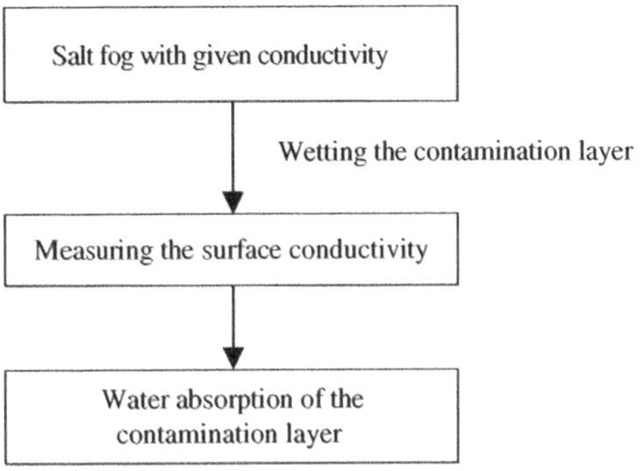

Figure 2.7 Measuring method for the saturated water absorption. *Source*: Cao et al. (2019)/with permission of IEEE.

contamination layer which is presented in Section III, the water mass attached to the insulator surface under different conditions can be obtained.

As shown in Figure 2.6, only one side (top surface) of the specimen was contaminated. Both the specimen that is made of glass and the contamination layer are hydrophilic. When placed in the fog room, conductive water film may be formed on not only the contaminated surface of the specimen but also the other side (bottom surface). In that case, the surface conductivity of two sides is comparable. However, the LCR meter is intended to measure the resistance of the wetted contamination layer between two conductive tapes. To prevent the formation of conductive water film on the other side of the specimen during the wetting process, the bottom surface was coated with RTV. Due to the good hydrophobicity of RTV coating, only separate droplets can be formed on the RTV coated surface even if the water flux is very high, as illustrated in Figure 2.8. Therefore, the resistance of the RTV coated surface is very large and its influence on the measurement of the wetted contamination layer resistance can be ignored.

In this study, kaolin was used to simulate the insoluble components in the contamination. Following IEC 60507 and IEEE Std. 4, the solid layer method was adopted to produce uniform pollution layers before the experiments. The test procedure is as follows:

1. A desired mass of kaolin was obtained by a high-precision balance. Then, the kaolin was poured on the glass specimen. After that, a certain amount of deionized water was sprayed on the glass surface. Finally, the contamination layer was coated evenly by a small brush and dried;
2. Place the specimen in a drying oven to dry the contamination layer;
3. Put the contaminated specimen into the fog chamber;
4. Prepare the NaCl solution with a conductivity of 9.02 mS/cm at 20 °C and then put it into an ultrasonic humidifier to generate a salt fog;
5. Measure the surface conductivity by the LCR meter during the wetting process.

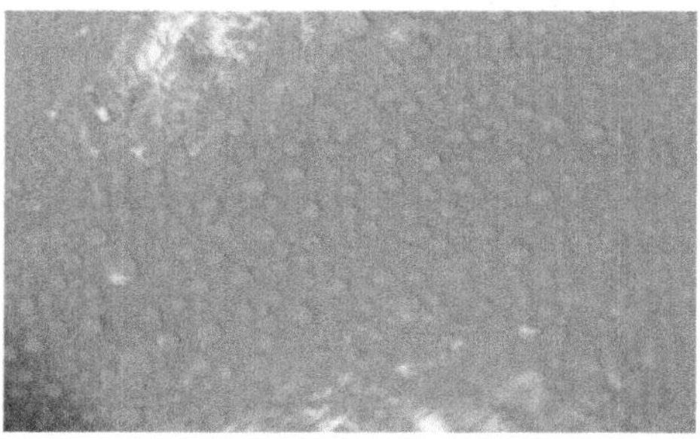

Figure 2.8 The surface appearance with RTV coated.

To ensure test repeatability, during each test, eight specimens, as shown in Figure 2.4, were wetted at the same time and the change of surface resistance of the contamination layers was measured continuously and recorded. The eight pieces of specimens were divided into four groups. Each group contained two samples which were arranged in the same way. If the resistance difference of specimens in each group was greater than 10%, then, the test results of this group were abandoned.

2.2.2 Measuring Method

During the wetting process, the salt fog particles deposited on the surface of specimens continuously and formed a water film. As there is a fixed ratio of conductive particles to water molecules in the salt fog, the more the fog particles captured, the more conductive ions in the contamination layer and the higher the surface conductivity of insulator.

The relationship between the surface conductivity of contamination layer and the conductivity of fog is:

$$K = \sigma \times h \tag{2.1}$$

where K refers to the surface conductivity; h denotes the thickness of the water film during wetting and is the conductivity of the salt fog.

Therefore, the water film thickness could be calculated by:

$$h = \frac{K}{\sigma} \tag{2.2}$$

In addition, it is obvious that the amount of water absorbed by the contamination layer per unit area is:

$$m_{\text{water}} = \rho_{\text{water}} \times h \tag{2.3}$$

where m_{water} is in mg/cm^2; ρ_{water} refers to the density of water. Therefore, the relationship between the surface conductivity and water absorbed can be obtained, as shown in Figure 2.9.

During the wetting process, the surface conductivity of the contamination layer was measured by the LCR meter. The temperature correction was carried out on the measured surface conductivity as follows:

$$K_{20} = k_t \cdot K_t \tag{2.4}$$

where K_{20} refers to the surface conductivity corrected under 20 °C; k_t means the surface conductivity measured at temperature t; and k_t is the temperature coefficient.

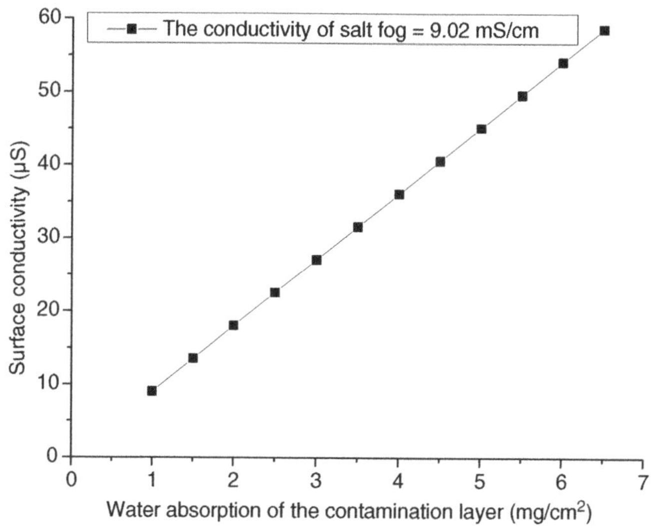

Figure 2.9 The relationships between the surface conductivity and water absorbed by the contamination layer. *Source*: Cao et al. (2019)/with permission of IEEE.

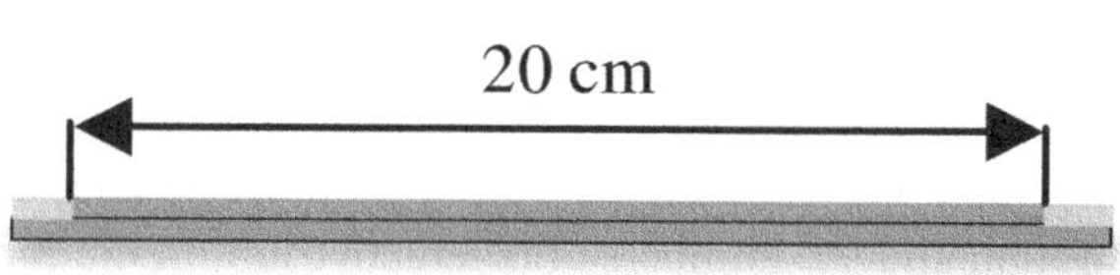

Figure 2.10 The schematic diagram of the contaminated hard rigid plastic specimen.

To validate the method proposed in this chapter, a very thin rectangular ultra-thin glossy transparent crystal hard rigid plastic, as illustrated in Figure 2.10, was chosen. It had a measurement area of 200 cm^2 (20 cm × 10 cm). After contamination, the plastic specimen was placed horizontally in the fog room. After four minutes of wetting, the specimen was taken out of the fog room. After that, the bottom surface of the plastic was dried by absorbent paper and then the LCR meter was used to measure the resistance of the wetted contamination layer. This test was repeated five times and an average water absorption was obtained.

The water absorption after four minutes of wetting was also measured by the weighting method. After contaminated the plastic following the method described in the test procedure, the plastic specimen was dried and weighted by a ME-T analytical balance with accuracy of 0.01 mg. Then, the specimen was placed horizontally in the fog room. After four minutes of wetting, the bottom surface and

Table 2.2 Measured results by two different methods.

Measuring method	Weight
Weighting method	8.12 mg/cm^2
Method proposed in this book	7.66 mg/cm^2

Note: The accuracy of the balance was 0.01 mg.

the conductive tapes were wiped by absorbent paper. Then, the specimen was weighted again. This weighting method was also repeated five times. The measured average weights by two methods are listed in Table 2.2. It can be seen that the difference between the weighting method and the method proposed in this chapter is 6.0%. Therefore, the proposed method is quite reliable.

2.3 Results

2.3.1 Trend of Surface Conductivity with Different Arrangement

The surface conductivity of the specimens over time with different arrangements in the wetting process is shown in Figure 2.11. In the wetting process, the salt fog particles in the fog room deposit on the surface of test specimen gradually and the water absorption and the numbers of conductive ions in the contamination layer are increased continuously at the same pace, which lead to the surface conductivity of the contamination layer increasing gradually. However, under the salt fog environment, the surface conductivity of the surface layer of the insulating surface will not increase all the time. With the increase of wetting time, the surface conductivity of the contamination layer increases first and then levels off, which means that the water absorption of the contamination layer is saturated.

As can be seen from Figure 2.11, compared with NSDD when the specimen was placed in an inclination angle of 45°, the arrangement of the contamination layers has a more significant influence on the moisture rate of the specimens. This is because, in the salt fog chamber, the wetting of the specimens is mainly caused by the capture of water droplets. Therefore, the moisture rate is affected by the arrangement of test specimen in the fog chamber. Due to the effect of gravity, the moisture rate on the top surface of inclined arrangement is higher than that on the bottom surface and also the vertical arrangement.

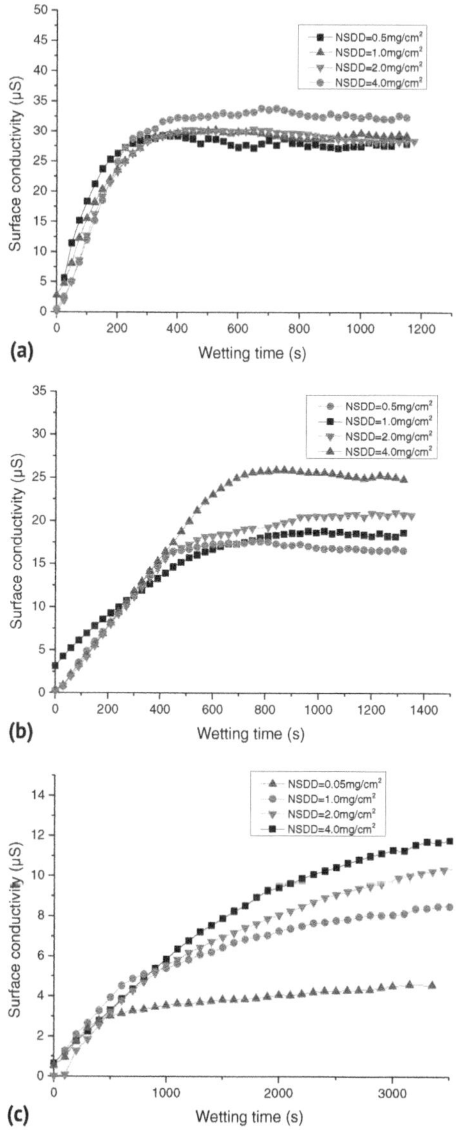

Figure 2.11 The surface conductivity over time with different arrangement: (a) contamination layer on the top surface (inclination angle of 45°); (b) contamination layer in vertical arrangement; (c) contamination layer on the bottom surface in horizontal arrangement. *Source*: Cao et al. (2019)/with permission of IEEE.

2.3.2 Effect of Contamination Layer Position (Top, Vertical, and Bottom)

2.3.2.1 The Top Surface

According to the method mentioned above, the saturated water absorption of the contamination layer with different inclination angles on the insulator surface could be calculated based on the surface conductivity during the wetting process. As shown in Figure 2.12, the saturated water absorption of the contamination layers is between 2.21 and 6.17 mg/cm^2, which is greatly affected by the inclination angle. The larger the inclination angle is, the less the water is attached to the specimen at saturated moisture.

Figure 2.13 shows the saturated water attached to the insulator surface with different NSDD when inclination angle of the contamination layer is 45°. The saturated water absorption of the insulating surface increases slightly with the increase of the NSDD. When NSDD increases from 0.5 to 4.0 mg/cm^2, the water attached to the insulator surface increases only less than 2.5%.

2.3.2.2 Vertical Contamination Layers

The water absorption of the contamination layer with the vertical arrangement is greatly influenced by the NSDD. As shown in Figure 2.14, when the NSDD of the

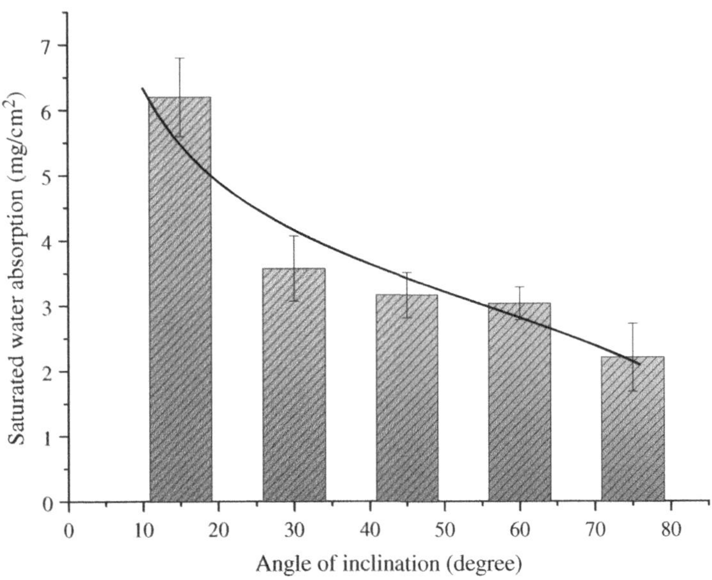

Figure 2.12 Saturated water absorption of the contamination layer with different inclination angle.

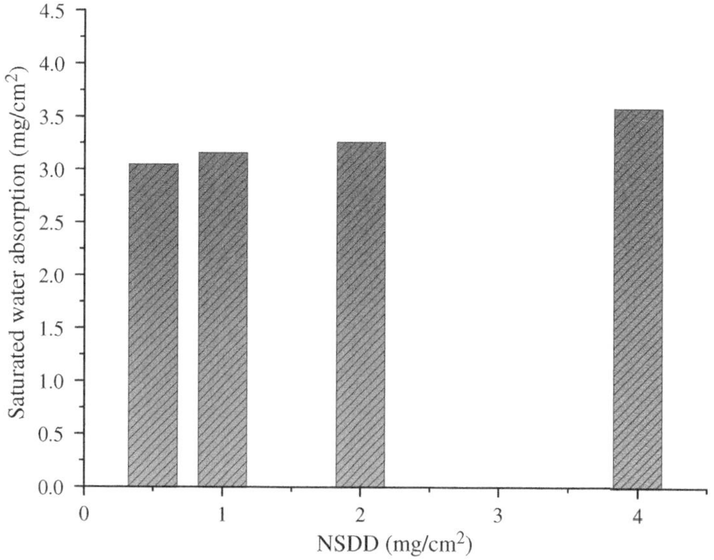

Figure 2.13 The saturated water attached to the insulator surface with different NSDD (inclination angle = 45°).

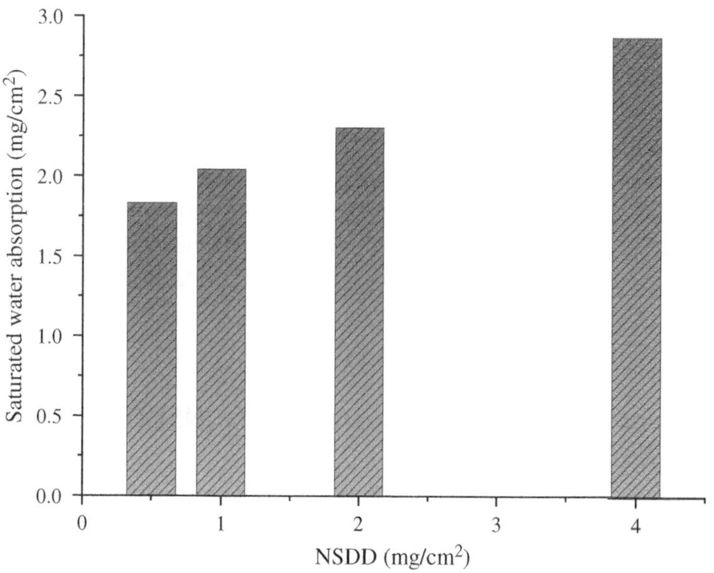

Figure 2.14 Saturated water attached to the insulator surface with vertical arrangement.

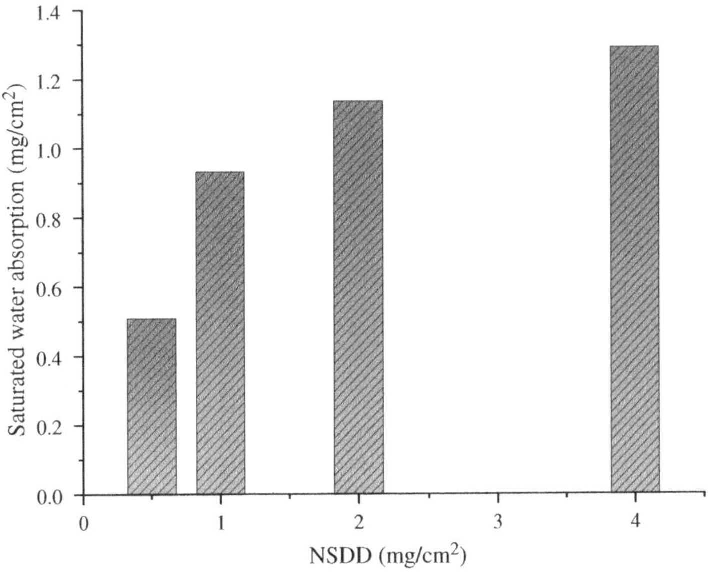

Figure 2.15 Saturated water attached to the insulator surface with different NSDD under the bottom surface.

specimens increases from 0.5 to 4.0 mg/cm^2, the water attached to the insulator surface increases as high as 56.7%.

2.3.3.3 On the Bottom Surface

The saturated water absorption of the contamination layer on the bottom surface is significantly lower than that on the top surface. Like the wetting process of the contamination layer with vertically arranged, the water attached to the contamination layer on the bottom surface increases significantly with NSDD, as shown in Figure 2.15.

2.4 Impact of Non-soluble Deposit Density (NSDD) and Surface Angle

The test results show that the saturated water absorption on the polluted insulator surface is affected by the inclination angle as well as NSDD. The larger the inclination angle of the contamination layers is, the less the water is attached to the insulator surface. With the increase of NSDD, the saturated water absorption of the contamination layer increases with the inclination angle.

2.4.1 Influence of NSDD on the Saturated Water Absorption

The insoluble substance, namely NSDD, in the contamination layers can increase the water attached to the insulator surface at saturated moisture condition. With the presence of insoluble substance, loose structures are formed on the specimen surface, as shown in Figure 2.16. These loose structures can store more water, which makes the saturated water absorption of the contamination layers increase with the increase of NSDD.

Compared with the contamination layer with vertical arrangement or on the bottom surface of inclined arrangement, the saturated water attached to the top surface of specimen with an inclination angle of 45° is hardly affected by the NSDD. Actually, this phenomenon was observed in the experiments when the specimen inclination angle was less than 60°. This can be explained by the fact that when the inclination angle is less than 60°, the saturated water absorption of the top surface is large. Therefore, after saturated moisture, the pores between insoluble substances are filled by the water attached to the specimen. In this case, the amount of water kept by the pollution layers is mainly determined by the inclination angle of the specimen.

2.4.2 Influence of the Surface Inclination Angle on the Saturated Water Absorption

For an inclined specimen with length of L and width of D, the salt fog particles fall on the specimen surface and gradually form a water film. As shown in Figure 2.17, when the water film flows down along the specimen surface and mixes with the falling fog particles, the thickness of the water film continuously increases.

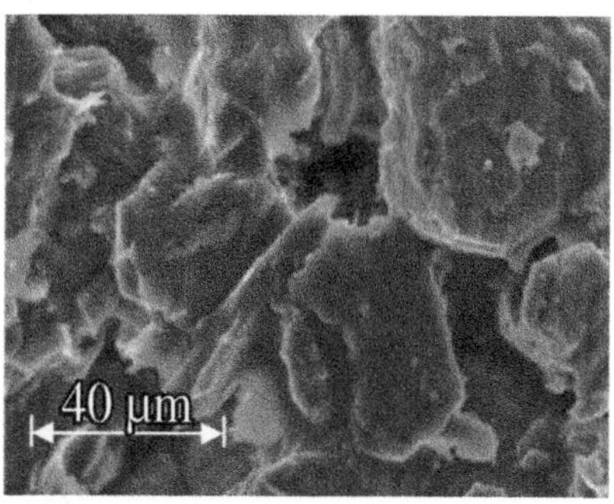

Figure 2.16 Contamination layer of Kaolin observed by SEM.

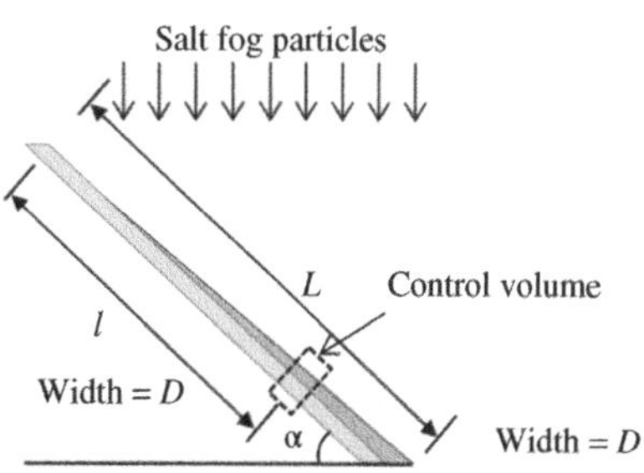

Figure 2.17 Wetting process of contamination layer on the top surface. *Source:* Cao et al. (2019)/with permission of IEEE.

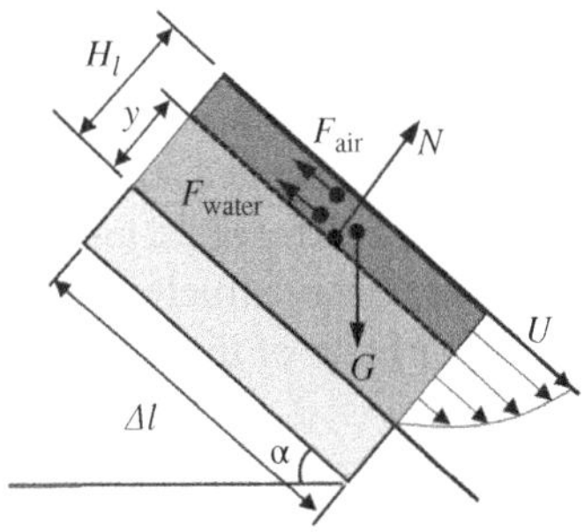

Figure 2.18 The force analysis of the water film element in the control volume. *Source*: Cao et al. (2019)/with permission of IEEE.

To investigate the saturated water absorption, a very small control volume in Figure 2.17 is taken to analyze the water film element as illustrated in Figure 2.18. The water film can be regarded as a laminar flow between two parallel plane plates, one of which is at rest while the other moves with velocity U. These plates are inclined at an angle α with regard to the horizontal plane. In the experiment, the movement of water film was very slow and a steady flow can be assumed.

For the water film element, Δl is very small, therefore, the thickness of water film is assumed constant and is denoted as H_l in the Figure 2.18. As the water film thickness is constant and the water film element has a width of D normal to the page, the pressure forces at the two end sections of the element will cancel each other. In addition, the viscosity of air (1.83×10^{-5} kg/m s) is about 1.83% of water (1.00×10^{-3} kg/m s) at room temperature (20°) and 1 atm pressure, therefore, the shear force Fair arising from the relative movement between the water film and air can be safely ignored. When applying the momentum equation to the dark blue part of the water film as shown in Figure 2.18, the following equation holds:

$$F_{\text{water}} - G \times \sin\alpha = 0 \tag{2.5}$$

where F_{water} is the shear force on the dark blue part, G is the weight of the dark blue part, and α is the inclined angle. F_{water} and G can be given by:

$$F_{\text{water}} = \tau_{\text{water}} \times \Delta A \tag{2.6}$$

$$G = (H_l - y) \times \Delta A \times \rho_{\text{water}} \times g \tag{2.7}$$

where τ_{water} is the shear stress of water, $\Delta A = \Delta l \times D$ is the surface area of the water film element.

Substituting Eqs. (2.6) and (2.7) to Eq. (2.5) yields:

$$\tau_{\text{water}} - (H_l - y) \times \rho_{\text{water}} \times g \times \sin\alpha = 0 \tag{2.8}$$

As water is a Newtonian fluid, the shear stress τ_{water} can be substituted by:

$$\tau_{\text{water}} = \mu_{\text{water}}\frac{du}{dy} \tag{2.9}$$

where u is the velocity of the water film at height y above the boundary. Thus,

$$\mu_{\text{water}}\frac{du}{dy} = (H_l - y) \times \rho_{\text{water}} \times g \times \sin\alpha \tag{2.10}$$

Integrating the above equation gives:

$$u = \frac{\rho_{\text{water}} \times g}{2\mu_{\text{water}}} \times \sin\alpha \times y \times (2H - y) + C \tag{2.11}$$

According to the no-slip condition, $u = 0$ when $y = 0$. Therefore, the constant C equals zero.

With the velocity u known, the volume flow rate Φ_l of water film at distance l from the top of the specimen can be obtained by integrating the velocity:

$$\Phi_l = \int_0^H u \times D \times dy = \frac{\rho_{\text{water}} \times g \times D}{3\mu_{\text{water}}} \times \sin\alpha \times H_l^3 \tag{2.12}$$

As the water film passing through the control volume all comes from the salt fog particles impinged on the specimen surface, from the top of the specimen to a distance l along the specimen surface, volume flow rate Φ equals the particle capture rate $\Phi_{\text{impinging}}$:

$$\Phi_l = \Phi_{\text{impinging}} = l \times D \times \cos\alpha \times s \tag{2.13}$$

where s is the sedimentation rate of salt fog per unit area. According to Eqs. (2.12) and (2.13), the thickness of water film element H_l can be given by;

$$H_l = \left(\frac{3\mu_{\text{water}} \times l \times s}{\rho_{\text{water}} \times g \times \tan\alpha}\right)^{\frac{1}{3}} \tag{2.14}$$

As indicated in Figure 2.2, the saturated water absorption was measured in the section between 0.9 L (18 cm) and L (20 cm). According to Eq. (2.14), the thickness ratio of water film at 0.9L and L is:

$$\frac{H_{0.9L}}{H_L} = (0.9)^{\frac{1}{3}} = 0.966 \tag{2.15}$$

As mentioned above, the thickness of the water film flowing down the specimen surface continuously increases. That means, in the measurement section, the thickness of water film was increased from $0.966H_L$ to H_L. Thus, it is fairly safe to say that the thickness was uniform and the measurement method, which was based on Eq. (2.3), was therefore valid.

Based on Eq. (2.14), the saturated water absorption M (mg/cm^2) in the section between $0.9L$ and L can be given by:

$$M = \frac{\int_{0.9L}^{L} \rho_{\text{water}} \times H_l \times D \times dl}{0.1 \times L \times D} = \frac{\int_{0.9L}^{L} \rho_{\text{water}} \times \left(\frac{3\mu_{\text{water}} \times l \times s}{\rho_{\text{water}} \times g \times \tan\alpha}\right)^{\frac{1}{3}} \times D \times dl}{0.1 \times L \times D} = 1.418 \times \left(\frac{\rho_{\text{water}}^2 \times \mu_{\text{water}} \times s \times L}{g \times \tan\alpha}\right)^{\frac{1}{3}} \tag{2.16}$$

Based on Eq. (2.16), it is found that there is a negative exponential power relationship between the water attached to the contamination layer and the tangent value of inclination angle. This conclusion is also confirmed by the experimental results. By applying the regression method to the results in Figure 2.12, the relationship between the mass of saturated water and the inclination angle can be obtained as:

$$m_{\text{water}} = 3.42 \times (\tan\alpha)^{-0.356} \tag{2.17}$$

with the correlation coefficient $R^2 = 0.9113$.

2.5 Summary

This chapter proposed a method to analyze the water absorption of the insulator surface during the wetting process. The influences of NSDD and the inclination angle of the specimen on the saturated water absorption of the contamination layer surface were also studied. From the experimental results, the following conclusions can be drawn.

1. For the contamination layer on the top surface of specimen in inclined arrangement, as the saturated water absorption is large, the influence of NSDD on the saturated water absorption is insignificant.

2. When the contamination layers reach saturated moisture, the loss of the contamination solution and the deposition of the fog particles reached a dynamic balance. It was found by experiment and theoretical analysis that there was a negative exponential power relationship between the saturated water absorption of contamination layer and the tangent value of inclination angle.
3. Compared with the contamination layer on the top surface, the wetting rate of the contamination layer on the bottom surface or of vertical arrangement is much slower and the water absorption after the saturated moisture is far less than that on the top surface. For the vertical arrangement or bottom surface of inclined arrangement, the saturated water absorption of the contamination layer increases with the increase of NSDD.

Bibliography

Cao, B., Wang, L., and Yin, F. Measurement of saturated water absorption of the contamination layer deposited on insulator surface, IEEE Sens. J., 2019, 19 (22), 10804–10811.

Cengel, Y.A. and Cimbala, J.M. Fluid Mechanics Fundamentals and Applications, McGraw-Hill, 2006.

Chrzan, K.L., Stec, C., and Farzaneh, M. Probes for spot measurement of surface conductivity on polluted insulators, IET Sci. Meas. Technol., 2007, 1, 295–300.

Deb, S., Choudhury, N.R., Ghosh, R. et al. Short time modified Hilbert transform-aided sparse representation for sensing of overhead line insulator contamination, IEEE Sens. J., 2018, 18 (19), 8125–8132.

Farzaneh, and Chisholm, W. Insulators for Icing and Polluted Environments, Wiley, 2009.

Gautam, B., Mizuno, Y., Matsubayashi, G. et al. Effect of wetting conditions on the contamination flashover voltages of polymer insulators, IEEE Proc. 8th ICPADM, Bali, 2006, pp. 538–541.

Huang, D., Li, X., Xu, T. et al. Saturated absorption of water quantity on the pollution layer of suspension insulator surface in steam fog, High Volt., 2014, 40 (11), 3349–3356.

Huang, X., Xie, C., and Li, H. Equivalent salt deposit density optical fiber sensor for transmission lines in power grid, IEEE Sens. J., 2017, 17 (1), 91–99.

IEC 60507-2013, Artificial Pollution Tests on High Voltage Insulators to Be Used on AC systems, 2013.

IEC TS 60815-1, Selection and Dimensioning of High-voltage Insulators Intended for Use in Polluted Conditions - Part 1: Definitions, Information and General Principles, 2008.

IEEE Std. 4-2013, IEEE Standard Techniques for High-Voltage Testing, IEEE Publication, 2013.

Lampe, W., Hoglund, T., Nellis, C. et al. Long-term tests of HVDC insulators under natural pollution conditions at the Big Eddy Test Center, IEEE Trans. Power Del., 1989, 4 (1), 248–259.

Leclerc, M., Bouchard, R.P., Gervais, Y. et al. Wetting processes on a contaminated insulator surface, IEEE Trans. Power App. Syst., 1982, 101 (5), 1005–1011.

Lin, F. Discussion of measured method for ESDD of pollution insulator, Insul. Surge Arrest., 1979, 40 (3), 24–31.

Mei, H., Wang, L., Guan, Z. et al. Influence of sugar as a contaminant on outdoor insulation characteristics of insulators in a substation, IEEE Trans. Dielectr. Electr. Insul., 2012, 19 (4), 1318–1324.

Obenaus, F. Fremdschichtueberschlag und Kriechweglaenge, Disch. Elektrotechnik, 1958, 12 (4), 135–136.

Wang, J., Wang, K., Zhou, M. et al. The natural contamination of XP-70 insulators in Shenzhen, China, IEEE Trans. Dielectr. Electr. Insul., 2016, 23 (1), 349–358.

Wang, L., Cao, B., Mei, H. et al. Effects of natural contamination components on the surface conductivity under saturated moisture, IEEE Trans. Dielectr. Electr. Insul., 2017, 24 (5), 2945–2951.

Werneck, M.M., dos Santos, D.M., de Carvalho, C.C. et al. Detection and monitoring of leakage currents in power transmission insulators, IEEE Sens. J., 2015, 15 (3), 1338–1346.

Yang, L., Zhang, F., Wang, Z. et al. Evaluation of wetting condition and its effects on pollution flashover voltage of aerodynamic insulators, IEEE Trans. Electr. Insul., 2016, 23 (5), 2875–2882.

Zhao, C., Mei, H., Zhu, M. et al. Development of contamination flashover pre-warning system and analysis of operating experience, IEEE Trans. Electr. Insul., 2015, 22 (4), 2234–2241.

3

Soluble Salt Dissolution in Contaminated Layers

3.1 Introduction

Chapter 2 analyzes the wetting characteristics of the contaminated insulation surface in the fog, and studies the relationship between the saturated water content of the pollution layer and the contamination condition. According to the F. Obenaus pollution flashover model, the insulator pollution flashover voltage mainly depends on the surface conductivity. The higher the surface conductivity, the lower the pollution flashover voltage. The surface conductivity is not only related to the amount of water attached to the insulator, but also to the amount of soluble salts dissolved at saturated moisture. The more contaminations on the surface of insulators are dissolved under the condition of saturated moisture, the more ions in the contamination solution and the higher the surface conductivity will be.

To evaluate the pollution flashover performance of insulators, artificial pollution flashover tests are carried out, and the concept of salt deposit density (SDD) is adopted in IEC 60815 to simulate the pollution severity of the natural contamination. Because of the high solubility, it is dissolved totally and forms the conductive particles in the wetting process. There is a significantly negative exponential relationship between the pollution flashover voltage and SDD. SDD is defined as the amount of salt (NaCl, a highly soluble salt) in the deposit on a given surface of the artificially polluted insulator (excluding the metal parts and assembling materials) divided by the area of this surface. In the present technical specification of artificial pollution test, the value of SDD is set to be the same as the measured Equivalent Salt Density (ESDD). By this specification, it can be found that, although its popularity, there is still a drawback of ESDD as the poor equivalence between artificial and natural contamination. The reason is as follows: the naturally pollutants deposited on the insulator surface are affected by the pollution sources in the nearby environment. It contains some highly soluble salts that dissolve readily into water such as NaCl, KCl, Na_2SO_4, etc., and slightly soluble salts like $CaSO_4$ that are hard to dissolve unless the water volume is very high. For naturally polluted insulators, the slightly soluble salts may be incompletely dissolved even under saturated moisture situation due to the limited water volume. However, in the ESDD measurement procedure, the slightly soluble salts are normally dissolved completely because plenty of water is used to collect the pollutants or clean the insulator surface. In this case, if the SDD, which equals the measured ESDD, is used to simulate the pollution severity in the artificial test, the flashover voltage obtained may diverge from that of naturally contaminated condition. Based on the above considerations, IEC 60815.4 also required the correction of soluble salt components in contamination during the selection and design of insulators in DC system.

The purpose of analyzing and correcting the soluble salt components (the proportions of highly soluble and slightly soluble salts) of the contamination accumulated on the insulator surface is to obtain a more accurate relationship between SDD of artificial contaminated insulators and ESDD of naturally contaminated insulators. Currently, the common methods used to analyze the contamination components are quite complicated. Generally, the cation anion charge balance, an indirect method, is used. However, the balance method exists only when all the major ion concentrations are known. The atomic absorption spectrophotometry is used to analyze Na^+, K^+, Mg^{2+}, etc., the ICP-atomic emission spectrometry is used to analyze Na^+, K^+, Ca^{2+}, Cl^-, SO_4^{2-}, NO_3^-, etc., and the ion chromatography is used to analyze Na^+, K^+, Ca^{2+}, Mg^{2+}. The discrepancy caused by different analyzing methods, and the inconsistent data units adopted by different research institutions result in poor data commonality. Moreover, these methods are quite expensive and inconvenient for on-site use. Due to the concerns mentioned above, the main objective of this chapter is to propose a method to evaluate and correct the soluble salt components of the naturally contaminated insulator. To achieve these, the dissolution amounts of soluble salts during ESDD measurement and under the saturated moisture condition were first studied. It will be particularly useful for the selection of insulators used in transmission lines and can give guidance on the selection of SDD value in the artificial pollution tests.

3.2 Experimental Arrangement

3.2.1 Dissolution of Soluble Salts During ESDD Measurement

The swab technique and washing technique (cap and pin insulators) are recommended for measurement of ESDD on insulator in IEC 60815 and IEEE Std. 4. First, the pollutants on the insulator surface should be either wiped off by absorbent cotton and collected in distilled water, or washed directly by the distilled water. Then, the electrical conductivity of the contamination solution, with all salts completely dissolved, is measured and converted to the value at temperature of 20 °C. After that, the ESDD on the insulator surface is calculated by electrical conductivity, water volume, and area of cleaned insulator surface. To facilitate site measurement, the volume of distilled water used for measuring ESDD is 100–300 mL (or more if required) by swab technique, and 500–1000 mL (or more if required) by washing technique. In China, the swab technique is more popular and the distilled water volume is determined by the following guidance: for IEC standard insulator (U70B/146), the volume of distilled water is 300 mL; for other insulators, the volume of distilled water varies is based on their surface area. The recommended volumes of distilled water for various surface areas are listed in Table 3.1. From Table 3.1, it can be seen that the water consumption is about 0.2 mL/cm^2.

The solubility of NaCl is very high. A volume of 100 mL water can dissolve 35.6 g of NaCl at 20 °C, while the ESDD of naturally contaminated insulator rarely exceeds 0.5 mg/cm^2. Therefore, it can be safely assumed that NaCl is completely dissolved during the ESDD measurement.

However, $CaSO_4$ is not easy to dissolve. Figure 3.1 shows the relationship between the electrical conductivity and mass of $CaSO_4{\cdot}2H_2O$ in 100 mL deionized water at 20 °C. It can be seen that the electrical conductivity first increases with the increase of $CaSO_4{\cdot}2H_2O$ and then levels off. For a volume of 100 mL deionized water, it can only dissolve 0.255 g of $CaSO_4{\cdot}2H_2O$ at 20 °C and the corresponding electrical conductivity is 1,760 μS/cm.

The relationship between ESDD and electrical conductivity is as follows:

$$\text{ESDD} = \frac{S_a \cdot V}{A} = \frac{(5.7 \times \text{EC}_{20})^{1.03} \cdot V}{A} \tag{3.1}$$

where EC_{20} is the electrical conductivity at 20 °C in S/m, S_a is the salinity of the suspension in kg/m^3 or mg/cm^3, V is the volume of the suspension (cm^3), and

Table 3.1 Volumes of distilled water recommended in China for ESDD measurement of various insulator surface areas.

Surface area (cm^2)	Water volume (mL)
≤ 1,500	300
(1,500~2,000)	400
(2,000~2,500)	500
(2,500~3,000)	600

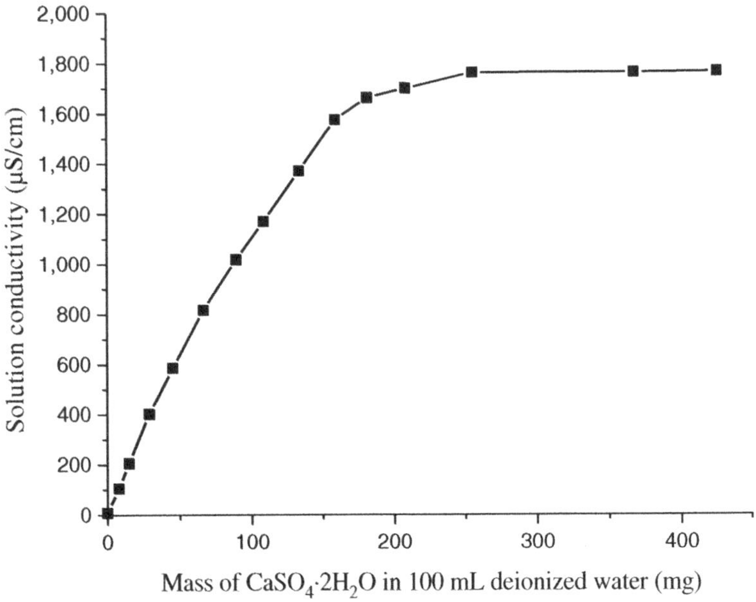

Figure 3.1 The relationship between electrical conductivity and mass of $CaSO_4 \cdot 2H_2O$ in 100 mL water. *Source*: Cao et al. (2019)/with permission of IEEE.

A is the area of the cleaned surface (cm^2). When the water volume used for ESDD measurement is 0.2 mL/cm^2, the maximum value of ESDD contributed by $CaSO_4 \cdot 2H_2O$ is 0.20 mg/cm^2 as calculated by Eq. (3.1). If the soluble salt in the contamination layer is only $CaSO_4$, it cannot be dissolved completely when the ESDD is greater than 0.20 mg/cm^2 ($CaSO_4 \cdot 2H_2O$ mass greater than 0.51 mg/cm^2). As ESDD of the naturally polluted insulators is generally less than 0.5 mg/cm^2, all the soluble salts are completely dissolved, and therefore, the dissolution of soluble salts during ESDD measurement is 100%.

According to Chapter 2, the saturated water absorption of contamination layer is 2–8 mg/cm^2, with the water film thickness of tens of micrometers. Considering the solubility of NaCl in water is 356 mg/mL and water density is about 1.0 g/mL at 20 °C, the NaCl in contamination will be completely dissolved under saturated moisture condition if its contribution on ESDD of the insulator does not exceed 0.7 mg/cm^2.

However, at 20 °C, the solubility of $CaSO_4 \cdot 2H_2O$ in water is 0.255 g/100 mL, and in the saturated moisture condition, the dissolution amount of $CaSO_4 \cdot 2H_2O$ is very small, which is only 5.1–20.4 $\mu g/cm^2$. Thus, $CaSO_4$ will not be completely dissolved during the insulator wetting process if its contribution on ESDD is greater than 4.6 $\mu g/cm^2$.

As the natural contamination consists of both slightly and highly soluble salts, the dissolution ratio of soluble salts k_s after the contamination been saturated wetted can be defined as

$$k_s = \frac{\text{ESDD}_{\text{sol}}}{\text{ESDD}} \tag{3.2}$$

where ESDD_{sol} is the salt dissolved under saturated moisture condition, and ESDD is the salt dissolved during the ESDD measurement procedure. ESDD_{sol}, due to its definition, is a more important parameter in studying the pollution flashover voltage of insulator than ESDD. As much more water is used during the ESDD measurement procedure, all slightly soluble salts are dissolved and ESDD is normally greater than ESDD_{sol}, leading to the value of k_s less than 1. Under saturated moisture condition, the ESDD_{sol} can be obtained by measuring the surface conductivity which is determined by the amount of soluble salt dissolved.

3.2.2 Test Specimens, Facilities, and Procedure

The device measuring surface conductivity of the contamination layer under saturated moisture is as shown in Figure 3.2. The size of test chamber is 50 cm × 80 cm × 100 cm.

The deionized water was turned into micron grade water mist by ultrasonic humidifier and conveyed into the test chamber, which made the humidity in the test chamber maintained at a high level. The test specimen was made of glass and the size was 20 cm × 30 cm × 0.5 cm. The surface conductivity of the test specimen was measured and recorded by RLC electric bridge.

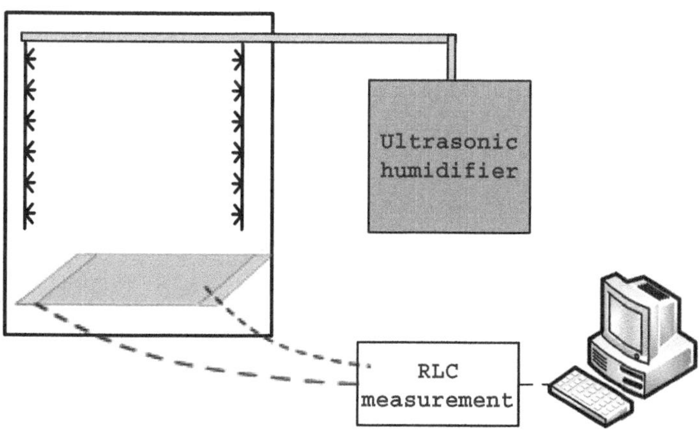

Figure 3.2 The device measuring the surface conductivity under the saturated moisture. *Source*: Wang et al. (2017)/with permission of IEEE.

Since the measuring voltage was low, the current through the contamination layer was very small and the thermal effect of the current on the contamination layer could be ignored. When the surface conductivity was stable, the water absorption of contamination layer reached saturation. The test specimen was placed in the test chamber horizontally during the test period. The surface of the test specimen was gradually moist and finally formed the visible water film. However, the pollution was not lost during the whole process.

The soluble salt of the contamination was simulated by NaCl and $CaSO_4$, and the insoluble substance is simulated by kaolin. The proportion of $CaSO_4$ in the soluble salt was determined by the contribution of $CaSO_4$ to the conductivity of the solution when ESDD was measured. By measuring the conductivity of $CaSO_4$ solution with different concentrations, we can find that the conductivity is directly proportional to the concentration of the $CaSO_4$ solution under the condition of non-saturation. In the same conductivity of $CaSO_4$ solution and NaCl solution at 20 °C, the mass ratio of the solute was 102.5:70. For example, the proportion of $CaSO_4$ in the test was 50%, which indicated that the contamination is scrubbed and dissolved into water, $CaSO_4$ and NaCl were equivalent to half of the conductivity of the solution. The pollution levels of NaCl and $CaSO_4$ during each experiment are shown in Table 3.2.

Table 3.2 The soluble salt of the contamination in artificial pollution test.

			ESDD (mg/cm^2)				
No.	**Proportion of $CaSO_4$**	**Soluble salt**	**0.025**	**0.05**	**0.1**	**0.15**	**0.2**
1	0%	$CaSO_4 \cdot 2H_2O$	0.0	0.0	0.0	0.0	0.0
		NaCl	17.5	35.0	70.0	105.0	140.0
2	25%	$CaSO_4 \cdot 2H_2O$	6.1	13.0	27.8	43.3	59.4
		NaCl	13.1	26.3	52.5	78.8	105.0
3	50%	$CaSO_4 \cdot 2H_2O$	8.8	17.5	35.0	52.5	70.0
		NaCl	12.1	25.9	55.5	86.7	118.9
4	75%	$CaSO_4 \cdot 2H_2O$	18.2	38.9	83.3	130.0	178.3
		NaCl	4.4	8.8	17.5	26.3	35.0
5	100%	$CaSO_4 \cdot 2H_2O$	24.2	51.8	111.0	173.3	237.7
		NaCl	0.0	0.0	0.0	0.0	0.0

Source: Cao et al. (2019)/with permission of IEEE.

In order to ensure the accuracy of the test results, the solid layer method was used to produce uniform pollution layers in this chapter. The quantitative soluble salts and kaolin were first weighed by high precision balance, and then poured on the glass plates. Afterward, some deionized water was dropped on it. Finally, the contamination layer was put evenly by a small brush and dried. After each test, the contamination would be scrubbed down and measured, if the difference between ESDD and its expected value was greater than 10%, the test results of this group would be abandoned.

3.3 Results

3.3.1 Highly Soluble Salts in Contamination

After placing the contaminated glass specimen into the damp test chamber, the contamination layers got damp and the soluble salts were dissolved. The surface conductivity of contaminated glass specimen was measured and recorded by the high precision electric bridge in wetting process in the test chamber. ESDD was 0.1 mg/cm^2, NSDD was 1.0 mg/cm^2 and the soluble salt of the contamination was NaCl. The change of surface conductivity during the wetting process in the test chamber was shown in Figure 3.3.

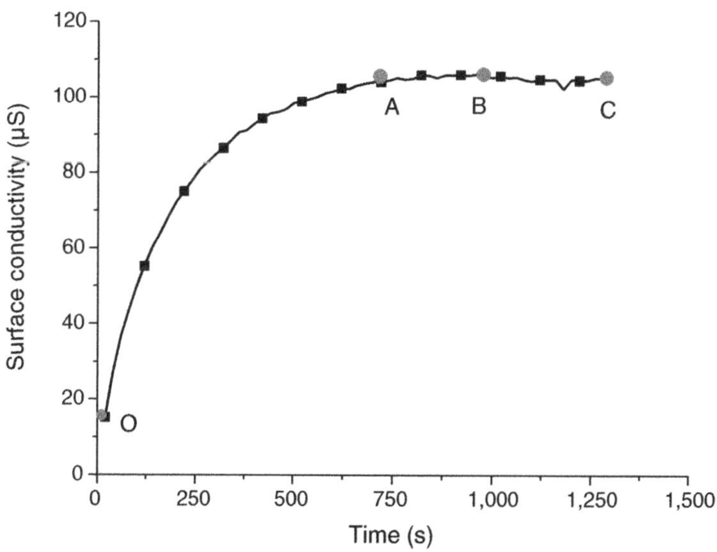

Figure 3.3 The tendency of surface conductivity during the wetting process. *Source:* Wang et al. (2017)/with permission of IEEE.

As is shown in Figure 3.3, the wetting process can be divided into two stages:

1. During the O–A stage, contamination on the glass specimen was gradually dissolved and the number of conductive particles in the contamination layer increased. The surface conductivity of the contamination layer increased gradually.
2. During the A–B stage, the soluble salt in the contamination was completely dissolved. The water absorption of the contamination layer still increased, but the number of conductive particles had reached the maximum value. Therefore, the surface conductivity of glass specimen did not change.

The glass surface began to appear visible water film at Point B, and the glass specimen was taken out at Point C.

It can be seen that the surface conductivity is related to the number of conductive particles in the contamination layer. The more conductive particles, the higher the surface conductivity is. If the soluble salts dissolved in the contamination were all dissolved, as shown in Figure 3.3, the surface conductivity was essentially unchanged, even if the water absorption of the contamination layer continuously increased.

The component of soluble salt in the contamination is NaCl. The surface conductivity of the glass specimens under the saturated moisture with different SDD is shown in Figure 3.4.

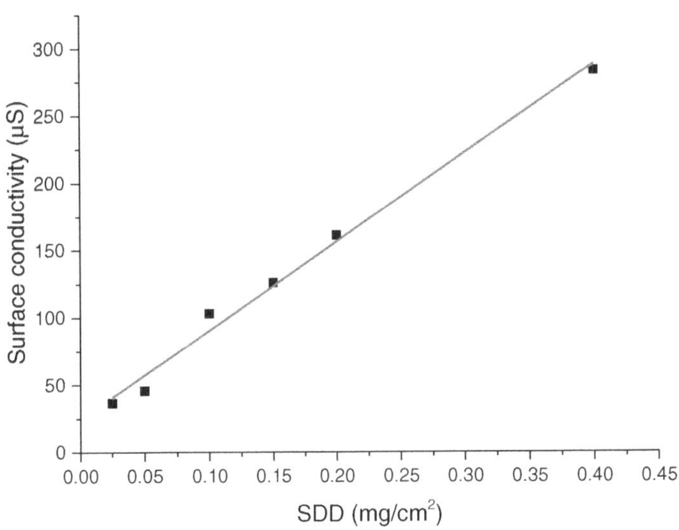

Figure 3.4 The relationship between the SDD and the conductivity of the glass surface under saturated moisture. *Source*: Wang et al. (2017)/with permission of IEEE.

According to the experimental results, it can be concluded that there was a proportional relationship between the surface conductivity of glass specimens under the saturated moisture and SDD.

If the component of soluble salt in the contamination is NaCl and SDD is 0.025–0.4 mg/cm^2.

$$K = 872 \times \text{SDD} + 10.1 \tag{3.3}$$

where K is the surface conductivity.

3.3.2 Slightly Soluble Salts in Contamination

The mixture of $CaSO_4$ and NaCl was utilized to simulate the soluble salt in the contamination, and the change trends of the conductivity of the surface are shown in Figure 3.5. The experimental results showed that the components of soluble salts had a great influence on the dissolution rate of contamination and the change trends of the surface conductivity of the glass specimen with the same ESDD varied with the proportion of $CaSO_4$. The higher proportion of $CaSO_4$ in the soluble salt, the slower the surface conductivity increases under the same moisture condition and the lower stable surface conductivity is.

The relationship among the surface conductivity, ESDD and the soluble salt component of the contamination is shown in Figure 3.6. The proportion of $CaSO_4$ in the soluble salt of the contamination is inversely proportional to the surface conductivity of the insulation surface. If the soluble salt in the

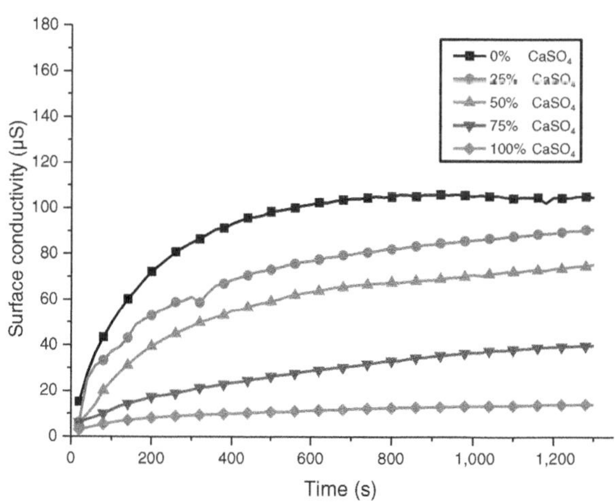

Figure 3.5 The surface conductivity of insulators with different proportions of $CaSO_4$ (ESDD = 0.1 mg/cm^2). *Source:* Cao et al. (2019)/with permission of IEEE.

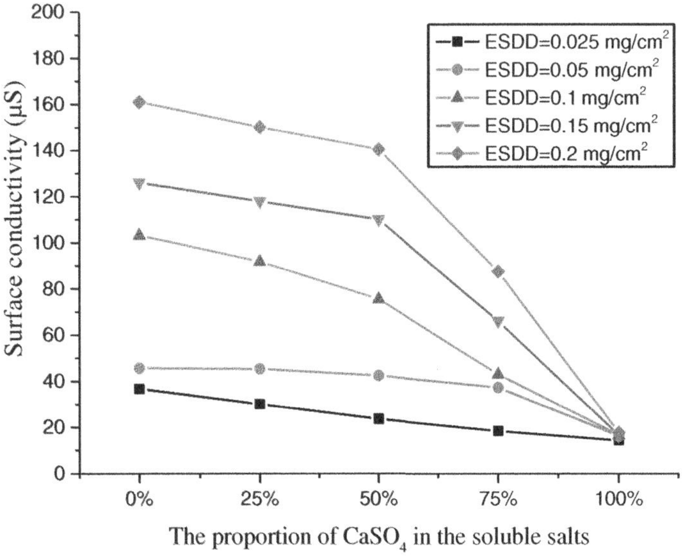

Figure 3.6 The relationship between the contamination component and the surface conductivity of the insulators under the saturated moisture. *Source*: Cao et al. (2019)/ with permission of IEEE.

contamination is single $CaSO_4$, the surface conductivity of the glass specimen is small, which is less related to ESDD.

3.4 Relationship Between ESDD and Conductivity

The experimental results above indicated that the surface conductivity of insulators varied with the contamination components. In the artificial pollution test, NaCl is generally utilized as the soluble salt. The solubility of NaCl is high and it is completely dissolved under saturated moisture. This is the main reason why there is a good correlation between the SDD and pollution flashover voltage besides the uneven distribution of contamination. Because of the different components of the natural contamination, the dissolved quantity of natural contamination is different, which leads to different surface conductivities of insulators and large dispersion of pollution flashover voltage.

In this section, the relationship among ESDD, contamination components and the surface conductivity under saturated moisture was studied. Also, the ratios of surface conductivity to ESDD were proposed as parameters to describe the components of natural contamination.

3.4.1 Highly Soluble Salts in Contamination

36.5 mg NaCl can be dissolved in 100 mg water at room temperature. Therefore, the soluble salts in the contamination were completely dissolved and formed the conductive particles in the saturated moisture condition. The relationship between the surface conductivity of insulators and SDD was obtained by the following derivation:

$$m_{\text{NaCl}} = \text{SDD} \times S \tag{3.4}$$

where S is the surface area of the insulators and m_{NaCl} is the quantity of NaCl on the surface of insulators.

The amount of NaCl (molecules) per unit area on the surface of insulators is

$$n_{\text{NaCl}} = \frac{m_{\text{NaCl}}}{W_{\text{NaCl}}} \tag{3.5}$$

where n_{NaCl} is the amount of NaCl (molecules) per unit area and W_{NaCl} is the relative molecular weight of NaCl.

Then, the molality of NaCl in the contamination layer under the saturated moisture is:

$$c_{\text{NaCl}} = \frac{n_{\text{NaCl}}}{V} = \frac{\text{SDD}}{W_{\text{NaCl}} \times \Delta h} \tag{3.6}$$

where c_{NaCl} is the amount of substance concentration of NaCl on the per unit area on the surface of insulators, V is the volume of the contamination solution and Δh is the thickness of the contamination solution.

The molar conductivity of electrolyte is the conductivity of the solution containing one mole electrolyte at a specific concentration. The molar conductivity of NaCl in the contamination solution of the insulator is:

$$\Lambda = \frac{\sigma}{c_{\text{NaCl}}} = \frac{\sigma \times W_{\text{NaCl}} \times \Delta h}{\text{SDD}} \tag{3.7}$$

where is conductivity of the contamination solution.

The relationship between conductivity of the contamination solution and SDD is:

$$\sigma = \frac{\Lambda \times \text{SDD}}{W_{\text{NaCl}} \times \Delta h} \tag{3.8}$$

Based on the relationship between the surface conductivity of the pollution layer and the conductivity of the contamination solution, the surface conductivity of pollution layer is

$$K = \Delta h \times \sigma \tag{3.9}$$

And the relationship between SDD and surface conductivity is:

$$K = \frac{\Lambda \times \text{SDD}}{W_{\text{NaCl}}} \tag{3.10}$$

The molar conductivity L of NaCl solution is related to the salinity of NaCl, as shown in Figure 3.7.

Based on the theory above, the relationship between the surface conductivity and the water absorption of contamination layer under the saturated moisture was analyzed. The water absorption mainly affected the salinity of the solution of the contamination layer. The relevant research results showed that the water absorption of contamination layer under the saturated moisture is generally between 5 mg/cm^2 and 10 mg/cm^2, in this range, the relationship between surface conductivity and SDD is as shown in Figure 3.8. The water absorption of the contamination layer had little influence on the surface conductivity of insulators. When the saturated water attached on the insulator surface is 3.0 mg/cm^2, the relationship between surface conductivity and SDD is shown in Figure 3.9.

The conductivity of contamination solution is affected by the interaction among ions, and the conductivity of NaCl solution is not strictly proportional to the concentration of NaCl in the entire solubility range. However, if the amount of soluble salts in the contamination is relatively small, then SDD is linear to the surface conductivity.

$$K = 987 \times \text{SDD} + 9.4 \tag{3.11}$$

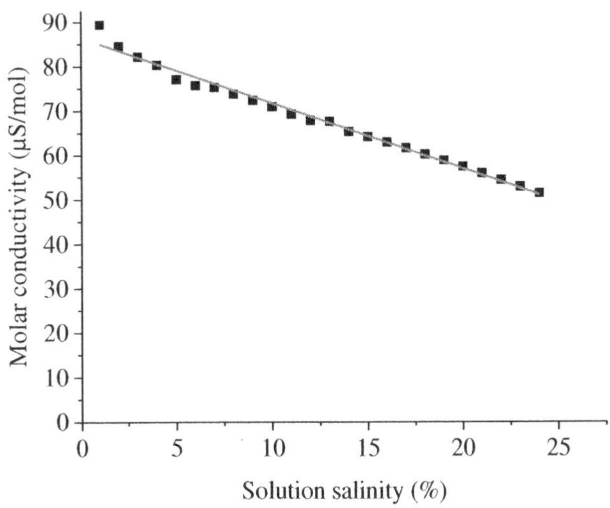

Figure 3.7 Relationship between salinity and molar conductivity of NaCl solution.

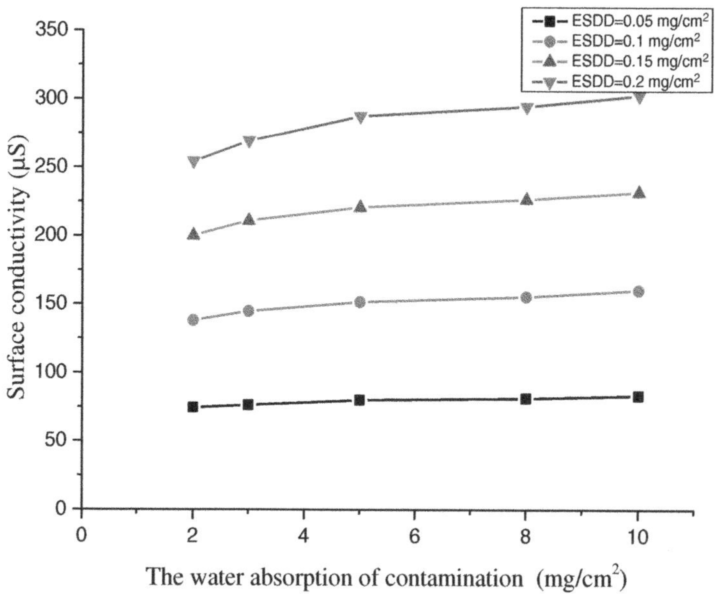

Figure 3.8 The relationship between surface conductivity and water absorption of contamination under the saturated moisture.

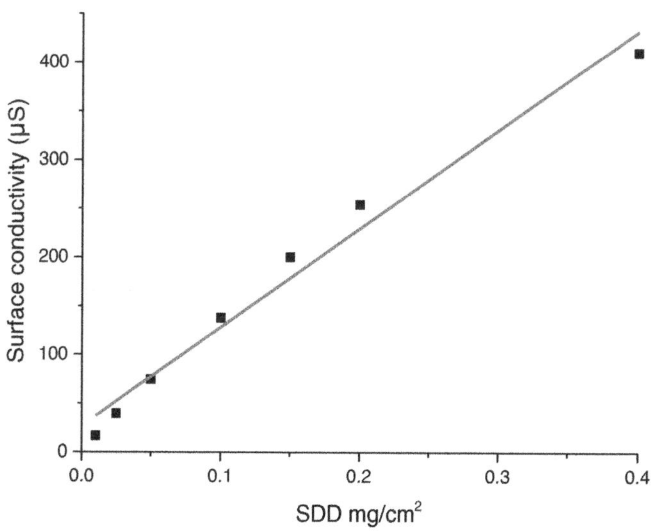

Figure 3.9 Relationship between the theoretical surface conductivity and SDD.

Although the SDD influences the ratio of surface conductivity to SDD, when SDD falls within the range of 0.025 to 0.4 mg/cm^2, the relationship between SDD and surface conductivity remains approximately linear. Furthermore, the discrepancy between the fitted and actual values of surface conductivity is insignificant, as demonstrated in Table 3.3.

3.4.2 Slightly Soluble Salts in Contamination

The natural contamination of insulators contains a certain proportion of slightly soluble salt ($CaSO_4$). The conductivity of $CaSO_4$ solution is approximately linear to the concentration.

$$\sigma = k_1 c_{\mathrm{CaSO_4}} \tag{3.12}$$

where $c_{\mathrm{CaSO_4}}$ is the concentration of $CaSO_4$ solution and k_1 is the ratio of solution conductivity to $CaSO_4$ concentration.

According to the mass ratio of the $CaSO_4$ to NaCl under the same conductivity of its solution at the standard temperature, the ESDD is converted to $CaSO_4$ mass density, and the $CaSO_4$ mass density on the surface of insulators can be expressed by:

$$\bar{\rho}_{\mathrm{CaSO4}} = k_m \mathrm{ESDD} \tag{3.13}$$

where ESDD is the equivalent SDD of natural contamination and km is the mass ratio of $CaSO_4$ to NaCl at the same conductivity of its solution in the standard temperature.

Table 3.3 Theoretical surface conductivity and SDD.

SDD (mg/cm^2)	Theoretical surface conductivity (mS)	Surface conductivity by linear fitting (mS)	Relative error
0.025	39.5	33.7	−14.7%
0.05	74.5	58.3	−21.8%
0.10	138.0	107.7	−22.0%
0.15	200.2	157.1	−21.5%
0.20	254.2	206.4	−18.8%
0.40	410.6	403.8	−1.7%

The concentration of $CaSO_4$ in the contamination solution under the saturated moisture is:

$$c_{CaSO_4} = \begin{cases} \dfrac{m_{CaSO_4}}{S\Delta h} = \dfrac{\bar{\rho}_{CaSO_4}}{\Delta h} & c_{CaSO_4} < c_{max} \\ c_{max} & c_{CaSO_4} \geq c_{max} \end{cases} \tag{3.14}$$

where c_{max} is the saturated concentration of $CaSO_4$; $CaSO_4$ is the mass of $CaSO_4$ in the contamination, and Δh is the thickness of the contamination solution.

According to Eq. (3.10), the surface conductivity of the insulators K can be

$$K = \begin{cases} \sigma\Delta h = k_1\bar{\rho}_{CaSO_4} & c_{CaSO_4} < c_{max} \\ \sigma_{max}\Delta h & c_{CaSO_4} \geq c_{max} \end{cases} \tag{3.15}$$

Because of the slight solubility, the $CaSO_4$ solution reaches saturation easily. If the soluble salt in the contamination is single $CaSO_4$, the surface conductivity is 3.68–18.7 μS under the saturated moisture. Due to the low solubility of $CaSO_4$, combined with the volumetric conductivity of saturated $CaSO_4$ solution, it can be inferred that the relationship between surface conductivity and ESDD is shown in Figure 3.10 when the attached water amount of the insulation surface fouling layer is 5.0 mg/cm^2.

In other words, if the soluble salt in the contamination is single $CaSO_4$ and ESDD is greater than 0.0037 mg/cm^2 then a part of $CaSO_4$ would not be

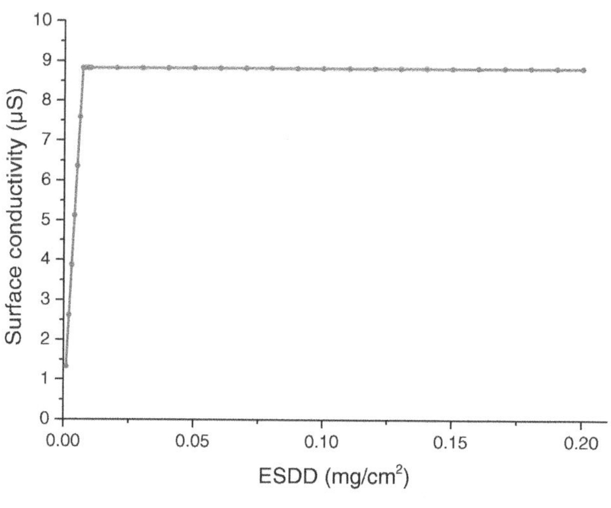

Figure 3.10 Relationship between the surface conductivity of insulator with calcium sulfate and ESDD (water absorption of contamination layer is 5 mg/cm^2).

dissolved in the wetting process and it has little effect on the external insulation performance of insulators.

3.4.3 Relationship Between the Surface Conductivity and ESDD

Because of different pollution sources in the different areas, the soluble salt of the natural contamination on the insulators is usually a mixture of NaCl and $CaSO_4$, etc. in different proportions. The soluble salts like NaCl can be dissolved completely, while slightly soluble salt like $CaSO_4$ is only partly dissolved. The surface conductivity is only related to the salt which is dissolved under the saturated moisture.

The relationship among ESDD, contamination components and the surface conductivity of the insulators under saturated moisture is shown in Figure 3.11.

If the proportion of $CaSO_4$ in the natural contamination remained unchanged, the surface conductivity of insulators would be positively correlated with ESDD and the proportion. And the larger the proportion of $CaSO_4$ is, the smaller the slope of the straight line is and the fitting parameters of straight lines are shown in Table 3.4.

The proportion of $CaSO_4$ in the contamination has little influence on the intercept *b* of the fitting line. However, the influence on the slope *k* is great.

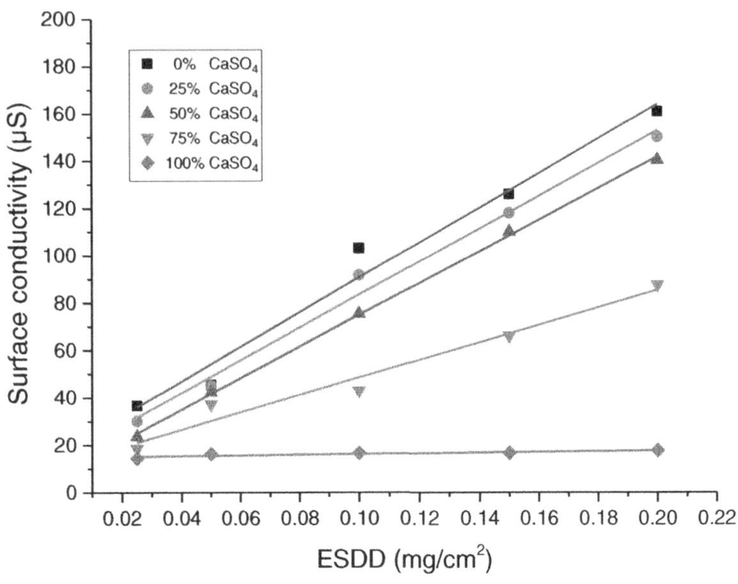

Figure 3.11 The surface conductivity of contamination under the saturated moisture.

Table 3.4 The coefficients of each relation curve between surface conductivity and ESDD.

Proportion of $CaSO_4$	k	b
0%	866	9.4
25%	763	9.9
50%	638	8.2
75%	368	10.7
100%	14	9.7

If the dissolution process of $CaSO_4$ and NaCl are independent, Slope $k_{x\%}$ can be deduced by

$$k_{x\%} = x\% \times k_{100\%} + (1 - x\%) \times k_{0\%} \tag{3.16}$$

where $k_{x\%}$ is the slope of the fitting line in the condition that the proportion of $CaSO_4$ is $x\%$; $k_{100\%}$ is the slope of the fitting line in the condition that the proportion of $CaSO_4$ is 100% and $k_{0\%}$ is the slope of the fitting line in the condition that the proportion of $CaSO_4$ is 0%.

However, the measurement results showed that slope k of the fitting line was not the weighted superposition of $k_{0\%}$ and $k_{100\%}$. For example, if the proportion of $CaSO_4$ in the soluble salt of the contamination was 50%, the slope $k_{50\%}$ would be 638, which was much larger than the slope deduced by formula (3.16), as shown in Figure 3.12. This indicated that the dissolution processing of NaCl and $CaSO_4$ in the wetting process is mutually affected.

The solubility of $CaSO_4$ in NaCl solution is as shown in Figure 3.13. According to Debye-Huckel theory, with the increase of NaCl concentration in the dilute solution, the ionic strength increases and its activity coefficient decreased. Therefore, the solubility of $CaSO_4$ in the NaCl solution is higher than that in the pure water. While in the high concentration NaCl solution, $CaSO_4$ formed the ion pair, which enhanced the short-range electrostatic interactions and reduced the solubility of $CaSO_4$. Assuming the saturated water absorption of contamination on the surface of insulators is 5 mg/cm^2. If ESDD was 0.025–0.4 mg/cm^2, the salinity of contamination solution would be in a range of 1–8%. In this concentration, the solubility of $CaSO_4$ was influenced by water salinity. The greater proportion of NaCl in the soluble salts is, the more $CaSO_4$ can be dissolved, which can be seen from Figure 3.13. The relationship among ESDD, components of natural contamination and surface conductivity is as shown in Figure 3.14.

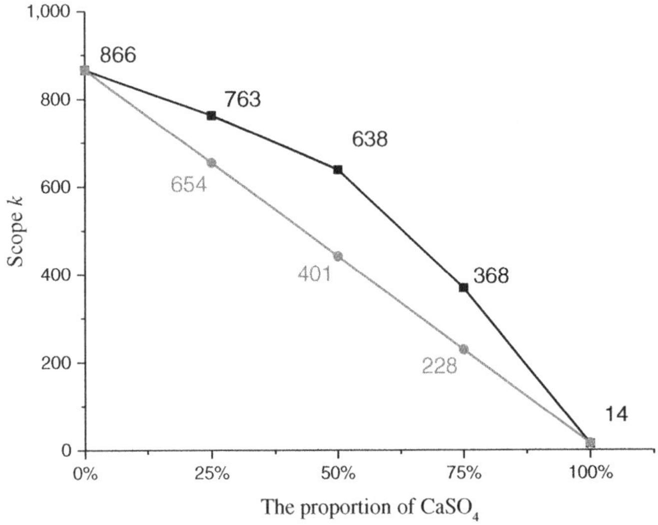

Figure 3.12 Coefficients of the relation curves between surface conductivity and ESDD. *Source*: Wang et al. (2017)/with permission of IEEE.

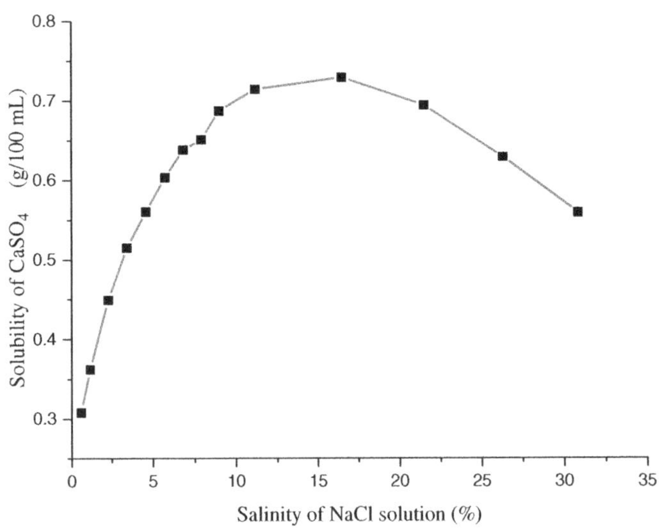

Figure 3.13 The solubility of $CaSO_4$ in NaCl solution. *Source*: Wang et al. (2017)/with permission of IEEE.

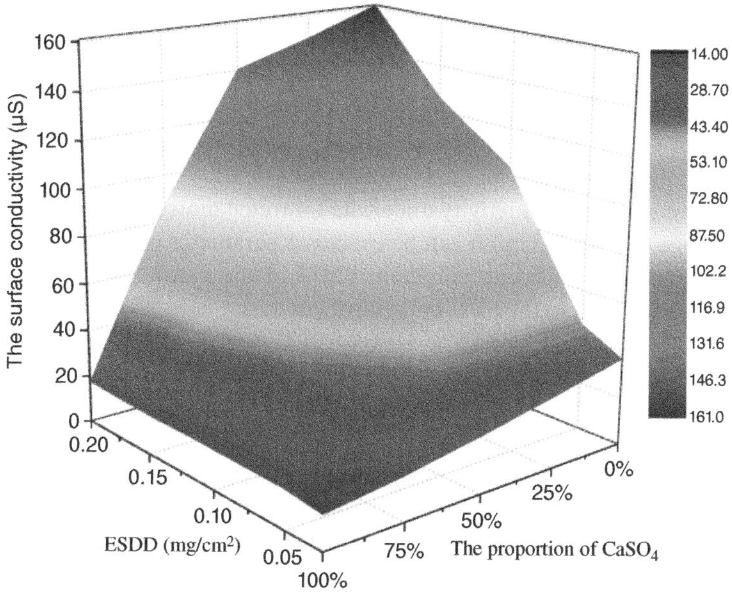

Figure 3.14 The relationship among ESDD, the contamination components and the surface conductivity. *Source:* Wang et al. (2017)/with permission of IEEE.

3.5 Summary

There are many slightly soluble salts such as $CaSO_4$ in the natural contamination of insulators, which make pollution flashover voltage of insulators have large scatter and lead to a lot of problems for the insulation design. In this chapter, the influence of contamination component to the surface conductivity of insulators was studied and the following conclusions were obtained:

1. The components of soluble salt in the natural contamination affected the external insulation characteristics of insulators. The surface conductivity reached a stable value at the saturated moisture. The higher proportion of $CaSO_4$ in the soluble salt, the slower the surface conductivity increased under the saturated moisture and the lower the stable surface conductivity would be.
2. NaCl utilized as the single soluble salt in the artificial contamination test was completely dissolved at saturated moisture and the surface conductivity increased linearly with the SDD. While for the natural contamination, the proportion of $CaSO_4$ in the soluble salt was inversely proportional to the

surface conductivity of the insulators. If the soluble salt in the contamination was single $CaSO_4$, the surface conductivity of the glass specimen was small, which had little relationship to ESDD.

3. There is a clear relationship among ESDD, contamination component and the surface conductivity of insulators. If two of the three parameters are measured, the remaining one can be calculated.
4. The ratio of surface conductivity under the saturated moisture to ESDD is in the range of 14–866, and it can be used as a parameter to describe the components of natural contamination. ESDD of the natural contamination can be deduced to the SDD in the artificial test.

Bibliography

Chisholm, W.A. Adapting satellite measurements of air pollution to select insulators, 2013 INMR World Congress, Vancouver, BC, Canada, 2013, pp. 208–244.

Chisholm, W.A. Selecting Insulators Using Satellite Measurements of Air Pollution, Insulators of INMR, June, 2016. [Online]. Available: http://www.inmr.com/selecting-insulators-using-satellite-measurements-air-pollution/

Chrzan, K.L., Vosloo, W.L., and Holtzhausen, J.P. Leakage current on porcelain and silicone insulators under sea or light industrial pollution, IEEE Trans. Power Del., 2011, 26 (3), 2051–2052.

Huang, D., Li, X., Xu, T. et al. Saturated absorption of water quantity on the pollution layer of suspension insulator surface in steam fog, High Vol. Eng., 2014, 40 (11), 3349–3356.

IEC TS 60815-1, Selection and Dimensioning of High-voltage Insulators Intended for Use in Polluted Conditions - Part 1: Definitions, Information and General Principles, 2008.

IEC TS 60815-4, Selection and Dimensioning of High-voltage Insulators Intended for Use in Polluted Conditions - Part 4: Insulators for D.C. Systems, 2016.

IEC TS 61245, Artificial Pollution Tests on High-voltage Ceramic and Glass Insulators to Be Used on D.C. Systems, 2015.

Jiang, X., Zhao, S., Xie, Y. et al. Study on fog flashover performance and fog-water conductivity correction coefficient for polluted insulators, IET Gener. Transm. Distrib., 2013, 7 (2), 145–153.

Jin, L., Tian, Z., Ai, J. et al. Condition evaluation of the contaminated insulators by visible light images assisted with infrared information, IEEE Trans. Instrum. Meas., 2018, 67 (6), 1349–1358.

Liljestrand, H.M. and Morgan, J. Error analysis applied to indirect methods for precipitation acidity, Tellus, 1973, 31, 421–431.

Oliveira, S.C. and Fontana, E. Optical detection of partial discharges on insulator strings of high-voltage transmission lines, IEEE Trans. Instrum. Meas., 2009, 58 (7), 2328–2334.

Qaddoumi, N.N., El-Hag, A.H., and Saker, Y. Outdoor insulators testing using artificial neural network-based near-field microwave technique, IEEE Trans. Instrum. Meas., 2014, 63 (2), 260–266.

Richards, C.S., Benner, C.L., Butler-Purry, K.L. et al. Electrical behavior of contaminated distribution insulators exposed to natural wetting, IEEE Trans. Power Del., 2003, 18 (2), 551–558.

Takasu, K., Shindo, T., and Arai, N. Natural contamination test of insulators with DC voltage energization at inland areas, IEEE Trans. Power Del., 1988, 3 (4), 1847–1853.

Taniguchi, Y., Arai, N., and Imano, Y. Natural contamination test of insulators at Noto testing station near Japan Sea, IEEE Trans. Power App. Syst., 1979, 98 (1), 239–245.

Wardman, J., Wilson, T., and Bodger, P. Volcanic ash contamination: limitations of the standard ESDD method for classifying pollution severity, IEEE Trans. Dielectr. Electric. Insul., 2013, 20 (2), 414–420.

Yang, Z., Jiang, X., Zhang, Z. et al. Study on the influence rules of soluble contaminants on flashover voltage of disc suspension insulators, IEEE Trans. Dielectr. Electric. Insul., 2016, 23 (6), 3523–3530.

Yuan, Q., Li, W., Zhou, S. et al. Integrated evaluation of aerosols during haze–fog episodes at on regional background in North China plain, Atmos. Res., 2015, 156, 102–110.

Zhang, Z., Jiang, X., Huang, H. et al. Study on the wetting process and its influencing factors of pollution deposited on different insulators based on leakage current, IEEE Trans. Power Del., 2013, 28 (2), 678–685.

Zhou, J., Xing, Z., and Deng, J. Characterizing and sourcing ambient PM2.5 over key emission regions in China I: water-soluble ions and carbonaceous fractions, Atmos. Environ., 2016, 135, 20–30.

4

Wetting Behavior of Insulator Contamination at Unsaturated Humidity

4.1 Introduction

As demonstrated in Chapters 2 and 3 of this book, the dissolution of contaminants on insulator surfaces is mainly governed by the water absorption of the pollution layer. In most regions, the ambient relative humidity remains below 100% for the majority of time. Under such conditions, the moisture uptake of the surface contamination layer does not reach saturation. This regime has received limited attention in previous studies, largely because pollution flashover events occur less frequently under unsaturated humidity conditions. However, in practice, such environmental conditions persist for much longer durations. Therefore, clarifying the wetting mechanisms under unsaturated humidity is of great significance for realizing long-term on-line monitoring of insulator contamination.

In this chapter, we focus on the wetting behavior of insulator contamination under conditions of unsaturated humidity, where the relative humidity remains below 100%. To investigate this, a controllable pollution test system was developed with adjustable temperature and humidity settings. In addition, a compact environmental chamber was designed to provide precise control over temperature, humidity, and pressure. Using these systems, a series of studies were conducted to explore how environmental parameters—including temperature, humidity—affect the wetting process. The influence of contamination characteristics, such as the presence of soluble salts, insoluble substances was also examined. Furthermore, the dynamics of water absorption of the insulator pollution layers at various humidities, was analyzed. These investigations provide a comprehensive understanding of the factors influencing wetting behavior in polluted insulators under unsaturated humidity conditions.

4.2 Experimental Arrangement

Reliable experimental methods and controlled environments are essential prerequisites for obtaining accurate test results. At present, there are few artificial pollution test systems that can precisely adjust both temperature and humidity. Although commercial temperature and humidity control equipment is available on the market, its integration with pollution test equipment is still limited. In order to study the influencing factors of insulator wetting characteristics, it is necessary to develop an economical, efficient, and reliable temperature and humidity control system.

Humidity generators generally operate on five fundamental principles:

1. Modification of the thermodynamic state of moist air: The state of a gas is determined by its pressure, temperature, and volume. If these remain constant, the moisture content remains unchanged. However, altering any of these parameters changes the moisture content. Humidity control based on thermodynamic P-V-T relationships can be achieved via the dual-pressure method, where pressure is varied; the dual-temperature method, where temperature is altered; or a combination of both.
2. Mixing methods, including converging flow and diverging flow techniques: In the converging flow method, moist or dry air is mixed with the target airstream to adjust humidity. In the diverging flow method, a stream of dry air is precisely split into two portions—one is saturated with water vapor while the other remains dry—then both are recombined to achieve the desired humidity.
3. Membrane permeation method: This approach uses membranes to exploit differences in water vapor partial pressures on either side. Water vapor diffuses through the membrane toward the lower pressure side, enabling humidity control.
4. Equilibrium vapor pressure method (Fixed-point temperature method): This technique leverages the principle that certain saturated salt solutions maintain a stable water vapor pressure under constant temperature and pressure, thereby providing a fixed relative humidity.
5. Chemical method: Based on the law of definite proportions, hydrogen and oxygen chemically react in the presence of a catalyst to generate a specific amount of water vapor, which can then be used for humidity control.

4.2.1 Two Methods for the Controlled Humidity in the Pollution Test

This section presents the design principles and operational features of two experimental systems. The first is a pollution test system that regulates temperature

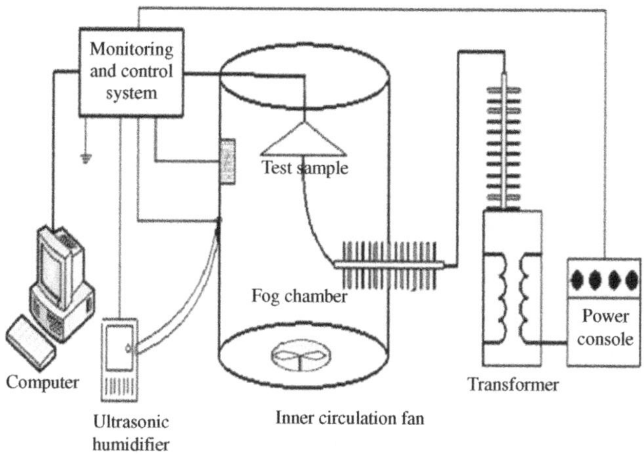

Figure 4.1 Structure diagram of temperature and humidity-controlled pollution test system.

and humidity using mixing methods. The second is a small-scale environmental chamber that enables precise control of temperature, humidity, and pressure through the equilibrium vapor pressure method. A comparative evaluation of the environmental control performance of both systems is also included.

The pollution test system with temperature and humidity control primarily comprises the following components: a small fog chamber with adjustable environmental conditions, a high-voltage power supply for pollution testing, a leakage current monitoring system, and dedicated control software. The overall system structure is shown in Figure 4.1.

The fog chamber of the system, is constructed from a cylindrical column of insulating acrylic glass, with a diameter of 85 cm and a height of 120 cm. It is equipped with an ultrasonic fogging unit, a dehumidifying exhaust system, internal circulation fans, evenly distributed air heaters along the inner wall, temperature and humidity sensors, and a real-time control software platform.

4.2.1.1 Pollution Test System with Controllable Temperature and Humidity

To achieve precise and stable control of temperature and humidity, the system dynamically adjusts hardware operation based on initial ambient conditions. Optimized control strategies are implemented through software to ensure minimal deviation from the target values. Typical control curves for humidity and temperature are shown in Figures 4.2 and 4.3, respectively.

The temperature and humidity values recorded by the sensors reflect only the conditions near the sensor probes. Uniformity across the chamber must be

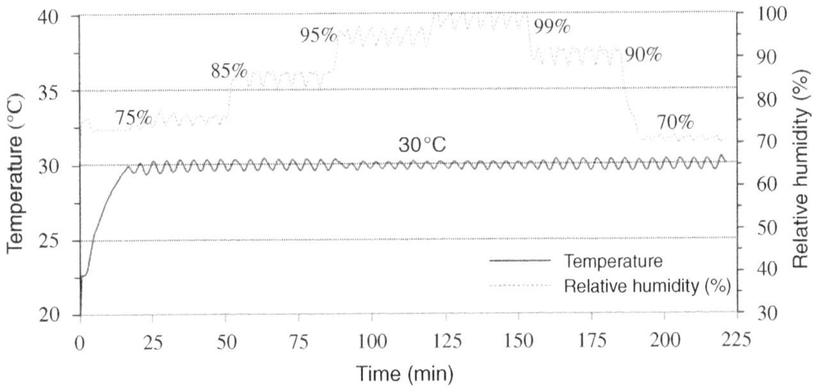

Figure 4.2 Effect of controlling indoor humidity with mist (target temperature remains unchanged at 30 °C).

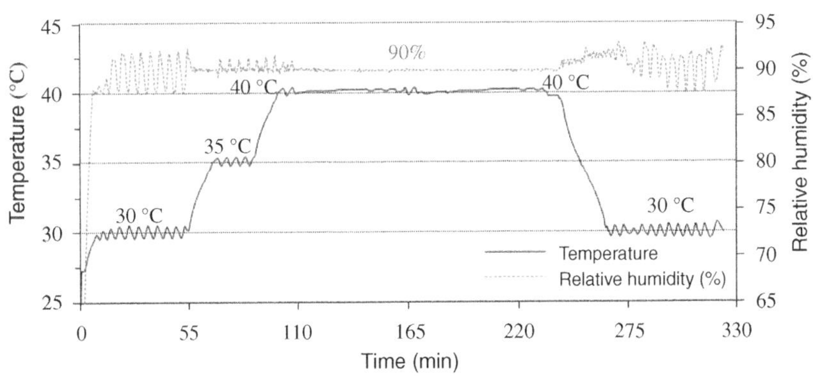

Figure 4.3 Effect of controlling indoor temperature with mist (target humidity remains unchanged at 90%).

verified experimentally. This is achieved by stabilizing the chamber at the target temperature and humidity, then taking measurements at various horizontal planes for no less than five minutes per position, while keeping the chamber sealed to prevent external disturbances. The results confirm that the temperature gradient is less than ±1 °C, and the humidity gradient is less than ±3%, both within the system's design tolerances. This also indicates that the vertical positioning of samples within the chamber has negligible influence on their wetting behavior.

Other components of the temperature and humidity-controlled pollution test system include:

1. Pollution test power supply: This consists of a voltage regulator, transformer, voltage divider, test control console, and measurement instruments. The transformer has a capacity of 100 kVA, with a maximum output voltage of 50 kV and a rated current of 2 A, suitable for insulator tests rated up to 35 kV.
2. Leakage current measurement system: This system includes a resistive contact-type leakage current sensor with built-in overvoltage and overcurrent protection, a data acquisition card, a power protection device, and a host computer. The sensor is connected in series between the insulator and ground, ensuring the entire leakage current passes through it. The analog signal is conditioned and digitized, then processed by bespoke software for real-time measurement, data storage, and analysis. A power protection device safeguards the computer against ground potential surges that may occur during flashover events. Measurement range: 0.5 mA–5 A; measurement accuracy: ±3%.
3. Monitoring and control software: Developed based on LabVIEW, the software provides real-time monitoring and analysis of test voltage, leakage current, and environmental conditions within the fog chamber.

This pollution test system, with its capability to regulate temperature and humidity, relies on relatively simple hardware and leverages virtual instrumentation for flexibility and scalability. It can maintain relative humidity from the ambient level up to 99% (±3% fluctuation), and temperature from ambient to ambient +30 °C, with an upper limit of 60 °C (±1 °C fluctuation). The system supports both standard pollution flashover testing and detailed studies of wetting behavior, making it a practical and economical platform for research in this field.

4.2.1.2 Small-scale Environmental System with Controlled Temperature, Humidity and Pressure

In Section 4.2.1, the temperature and humidity-controlled pollution testing system utilizes the converging flow method within its small fog chamber to precisely regulate humidity. Meanwhile, the environmental system with controlled temperature, humidity, and pressure, employs the equilibrium vapor pressure method for accurate humidity control. This system consists of a sealed container holding a saturated salt solution placed inside a temperature-controlled chamber. After sufficient time, a three-phase thermodynamic equilibrium is established among the salt, its solution, and the gas above it.

This dynamic equilibrium results from ionization and hydration within the saturated solution, and concurrent evaporation and condensation at the liquid surface. These processes stabilize the humidity in the vapor-liquid interface region. Over time, this humidity field expands upward until the entire sealed space reaches a uniform and stable humidity. The relative humidity of the microenvironment depends on the type of salt used; each salt yields a fixed humidity at a

Table 4.1 Reference table of relative humidity values of saturated salt solutions at different temperatures.

Salt	Temperature (°C)										
	5	10	15	20	25	30	35	40	50	60	70
$LiCl.xH_2O$	11%	11%	11%	11%	11%	11%	11%	11%	11%	--	--
KOH	14%	12%	11%	9%	8%	7%	7%	6%	6%	6%	--
CH_3COOK	--	23%	23%	23%	23%	22%	--	--	--	--	--
$MgCl_2.6H_2O$	34%	34%	33%	33%	33%	32%	32%	32%	31%	29%	--
$K_2CO_3.2H_2O$	43%	43%	43%	43%	43%	43%	--	--	--	--	--
$Mg(NO_3)_2.6H_2O$	59%	57%	56%	54%	53%	51%	50%	48%	45%	--	--
NaBr	64%	62%	61%	59%	58%	56%	55%	53%	51%	50%	--
KI	73%	72%	71%	70%	69%	68%	67%	66%	65%	63%	62%
NaCl	76%	76%	76%	75%	75%	75%	75%	75%	74%	75%	--
$(NH_4)_2SO_4$	82%	82%	82%	81%	81%	81%	80%	80%	79%	--	--
KCl	88%	87%	86%	85%	84%	84%	83%	82%	81%	80%	--
KNO_3	96%	96%	95%	95%	94%	92%	91%	89%	85%	--	--
K_2SO_4	99%	98%	98%	98%	97%	97%	97%	96%	96%	--	--
Purified distilled water	100%										

given temperature. By selecting different salts, a range of humidity levels can be achieved (see Table 4.1). Thus, the micro-environmental system comprises two main components: A sealed glass vessel containing the saturated salt solution and a temperature-controlled chamber (see Figure 4.4). By placing the sealed vessel inside the temperature-controlled chamber, a stable environment of controlled temperature, humidity, and pressure is formed. Adjustments can be made by replacing the salt solution to change humidity, modifying the chamber temperature (which slightly affects humidity, within acceptable limits), or using a vacuum pump to adjust the internal air pressure.

The temperature, humidity and air pressure control effects of the environmental system designed in this section are shown in Figures 4.5 and 4.6. Its features are as follows:

1. High precision: The use of saturated salt solutions allows for accurate and stable control of relative humidity, with minimal fluctuation after equilibrium is reached.

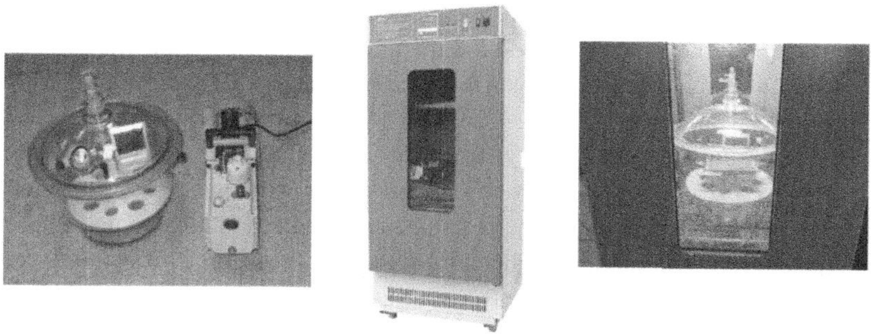

(a) (b) (c)

Figure 4.4 Physical picture of the small-scale environmental system: (a) glassware; vacuum flask; (b) thermostat; (c) glassware in a thermostat.

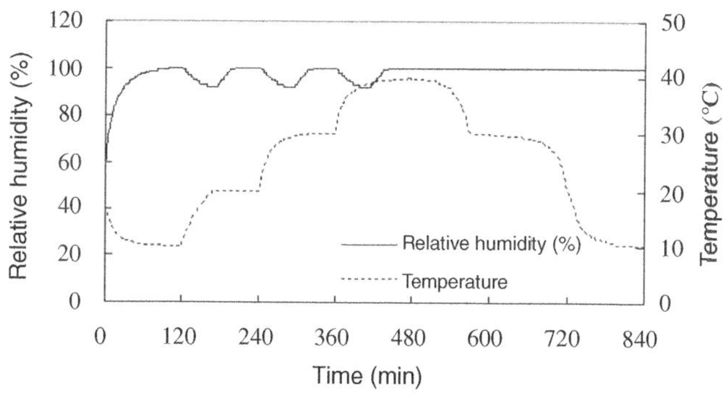

Figure 4.5 Control effect of target humidity 100% and continuous temperature adjustment (adjustment interval 120 minutes).

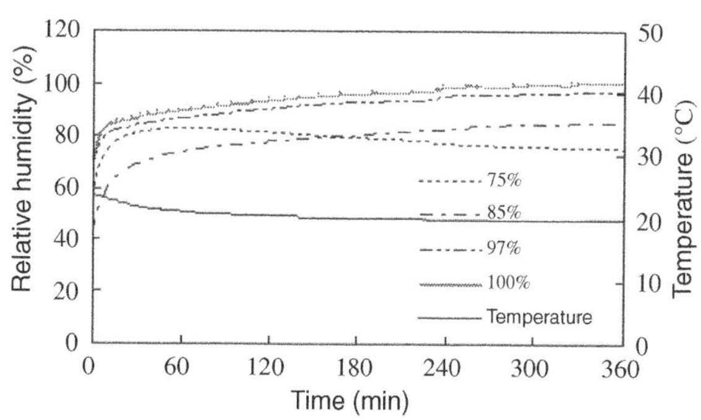

Figure 4.6 Control effect with a target temperature of 20 °C and varying humidity (saturated salt solution were replaced).

2. Adjustable temperature: If the selected salt solution maintains a relatively constant humidity across a temperature range, the system temperature can be continuously adjusted. However, the time required to reach thermal stability exceeds 60 minutes.
3. Limited humidity adjustability: Humidity cannot be adjusted continuously at a fixed temperature. Even with temperature variations, the change in humidity is limited (typically 1–5%). Large humidity adjustments require replacing the salt solution. The system typically requires over 3 hours for humidity stabilization. Temperature variations also affect the time required for humidity to stabilize.

4.2.1.3 Comparison Between Two Systems

Both the pollution test system and the environmental system can regulate temperature and humidity, but they differ significantly in performance and application:

1. Spatial capacity: The pollution test system features a larger chamber suitable for both sample and insulator testing while the environmental system is compact and only suitable for small sample testing.
2. Stabilization time: The pollution test system uses air heaters and the converging flow method for humidity control, achieving stability within 30 minutes. The environmental system by contrast, uses a thermostat and saturated salt solutions, requiring about one hour for temperature stabilization and more than three hours for humidity stabilization.
3. Control accuracy: The pollution system offers moderate control precision, with some post-stabilization fluctuation, while the environmental system offers high precision and minimal fluctuation.
4. Applicability to flashover testing: The pollution system supports pollution flashover testing while the environmental system does not.
5. Suitability for dynamic analysis: The pollution system, with shorter stabilization time, is better for studying transient effects. But the environmental system, while slower, more closely resembles natural humidity variation processes.

Therefore, this chapter combines both systems to analyze the influence of environmental factors on the wetting characteristics of insulators.

4.2.2 Test Specimens, Facilities, and Procedure

There are two primary methods used to characterize the wetting degree on insulator surfaces. One is an indirect method, which involves monitoring changes in surface conductivity after the contaminated insulator surface becomes moistened. The other is a direct method, which quantifies the amount of water absorbed by the insulator surface.

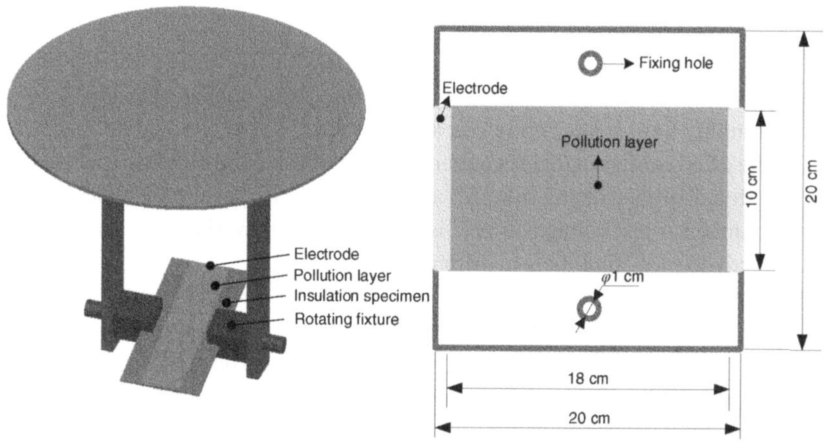

Figure 4.7 Schematic diagram of specimen mounting and fixture structure.

This section focuses on the effects of environmental parameters—specifically ambient temperature, relative humidity under unsaturated moisture conditions—on the wetting behavior and wetting process of insulators. The variation in surface conductivity during the wetting process is used to characterize the differences in wetting performance under various environmental conditions.

The specimen mounting method and its schematic are shown in Figure 4.7. The specimen consists of a glass plate with a thickness of 0.2 cm and dimensions of 20 cm × 20 cm. A 10 cm × 20 cm central area is coated with the artificial contamination layer. On both sides of the polluted area, there is a 1 cm diameter hole used for securing the specimen using a clamp. The clamp allows 360° rotation of the glass plate, enabling adjustments to the inclination angle of the specimen. Aluminum foil electrodes, each 1 cm in width, are affixed to both ends of the pollution-coated area to ensure uniform voltage distribution across the surface.

The test voltage is applied to the specimen through an AC isolation transformer and AC voltage regulator. The effective value of the test voltage should not be too low, otherwise, the surface leakage current during the wetting process will be minimal, adversely affecting the measurement accuracy of surface conductivity. Conversely, if the effective value is too high, the surface leakage current will become excessively large during wetting, leading to significant thermal drying effects caused by the current, which may interfere with the investigation of environmental influences. Considering these factors, and based on both theoretical analysis and experimental comparison, the effective test voltage is set at 50 V.

The soluble salt in the artificial contamination is NaCl, and the insoluble material is kaolin. The salt deposit density (SDD) is set at 0.1 mg/cm^2, and the non-soluble deposit density (NSDD) is set at 1.0 mg/cm^2. The experimental procedure is as follows:

1. The small fog chamber is regulated to the target temperature and relative humidity until stable conditions are achieved.
2. The dried contaminated specimen is placed horizontally in a fixture with the polluted surface facing upward, and then carefully positioned inside the fog chamber. A test voltage of 50 V is applied, and the variation in surface conductivity is continuously monitored.
3. If the surface conductivity exhibits only minor changes over a 30-minute period, it is considered that the specimen has reached full wetting under the given temperature and humidity conditions. The corresponding surface conductivity value is recorded to characterize the wetting behavior under these environmental parameters.
4. For each test point, six specimens are tested. The representative surface conductivity for that condition is defined as the average of the middle four values, after excluding the maximum and minimum readings among the six.
5. Since the temperature and humidity inside the small fog chamber can only be maintained at levels higher than the ambient conditions, the experiments must be conducted opportunistically, depending on favorable weather conditions.

4.3 Results

4.3.1 Effects of Environmental Humidity on the Steady-state Contamination Moisture

The effect of ambient humidity on leakage current is shown in Figure 4.8, with the same pollution degree, as the atmospheric humidity increases, the leakage current increases linearly. Under the same humidity, the leakage current also increases with the contamination degree. When the atmospheric humidity is lower than 75%, the equilibrium leakage currents are nearly zero. It is caused by the wetting characteristics of soluble salt in the contamination, because when NaCl, NaCl solution, and upper atmosphere form a three-phase thermodynamic equilibrium system at 25 °C, the atmospheric relative humidity of the air above the solution is 75%.

4.3.2 Influence of Insoluble Substances

The impact of insoluble components in pollution on wetting characteristics was primarily investigated through two approaches: (1) measuring the variation

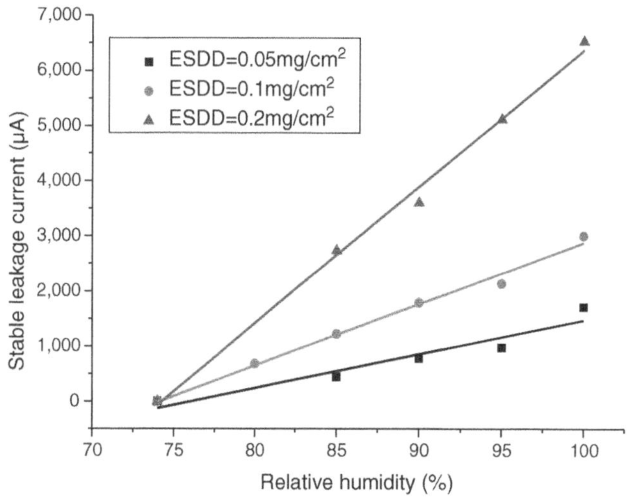

Figure 4.8 The relationship between equilibrium leakage current and atmospheric relative humidity. *Source*: Cao et al. (2022)/with permission of IEEE.

Figure 4.9 Glass specimens and test environment used for the water absorption experiment.

in surface conductivity of contaminated specimens containing five different types of insoluble substances under varying temperature and humidity conditions; (2) weighing the water absorption of a unit mass of each insoluble substances.

Surface conductivity tests were conducted in a small-scale environmental system with controllable temperature, humidity, and pressure. The SDD was fixed at 0.1 mg/cm^2 using NaCl, and the NSDD was 1.0 mg/cm^2. A constant voltage of 50 V was applied. As illustrated in Figure 4.9: In the water absorption test, specimens with dimensions of 10 cm × 10 cm × 0.09 cm were used. The SDD remained at 0.1 mg/cm², while the theoretical NSDD was set at 2.0 mg/cm²—actual values were determined by precise weighing.

The procedure for the water absorption measurement is as follows:

1. Clean specimens were dried in a thermostat at 50 °C and 10% relative humidity, and then weighed.
2. Pollution-coated specimens were dried under the same conditions and weighed.
3. The contaminated specimens were then placed in a controlled environment chamber at 25 °C with relative humidity levels of 85, 95, and 100%, respectively, and weighed again after moisture absorption.
4. The water absorption per unit mass of insoluble substances was calculated to characterize the wetting behavior of each material.

The influence of insoluble substances composition on surface conductivity and water absorption is presented in Figures 4.10 and 4.11. The experimental results indicate the following: (1) Surface conductivity of specimens coated with pollution containing different insoluble substances varies with humidity, generally increasing as relative humidity rises. (2) At low humidity levels, specimens coated with coal powder exhibit the lowest surface conductivity, while those with charcoal powder show the highest. This trend persists under high humidity conditions, with the difference between them becoming more pronounced. The water absorption test results are consistent: charcoal powder absorbed the most moisture, while diatomite and coal powder absorbed the least.

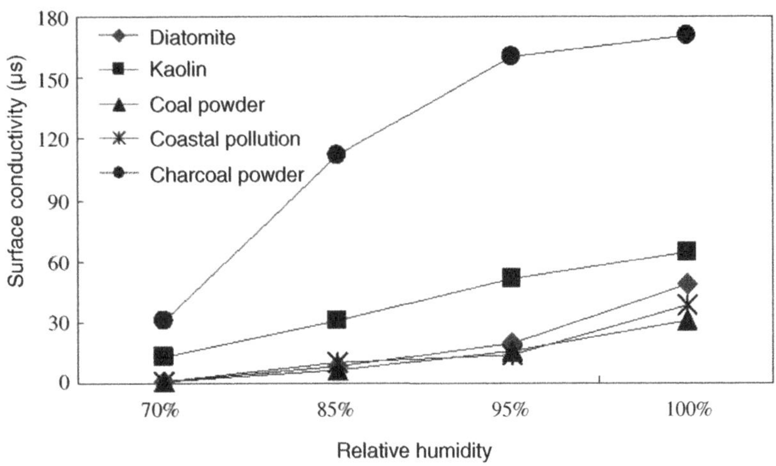

Figure 4.10 Influence of different insoluble substances on surface conductivity (at 25 °C).

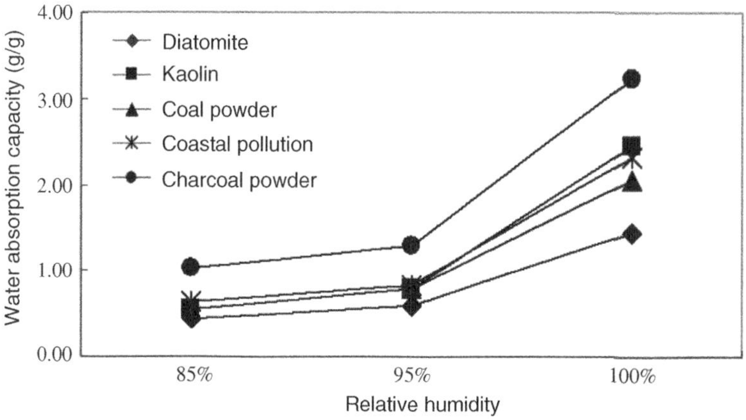

Figure 4.11 Water absorption characteristics of different insoluble substances (at 25 °C).

Table 4.2 Specific surface area of five insoluble substances.

Insoluble substances	Specific surface area (m^2/g)	Total pore volume (cm^3/g)	Average pore diameter (nm)
Diatomite	13.6978	0.037444	10.93441
Kaolin	19.3335	0.053955	11.16302
Coal powder	13.2129	0.035672	10.79925
Charcoal powder	78.7288	0.086115	4.37526
Natural coastal contamination	19.9592	0.069020	13.83221

The specific surface area and microscopic morphology of the five insoluble substances are shown in Table 4.2 and Figure 4.12.

Based on the experimental data, the following analysis can be made: Charcoal powder has the largest specific surface area, while coal powder and diatomite have the smallest. Under identical environmental conditions, materials with larger surface areas tend to absorb more water, resulting in higher surface conductivity. Therefore, the effect of insoluble material composition on wetting behavior is closely related to its water absorption capacity, which in turn depends on its specific surface area.

The influence of insoluble material content was further studied using kaolin as a representative material. Results are shown in Figure 4.13. The following observations were made: Under unsaturated conditions, higher NSDD leads to

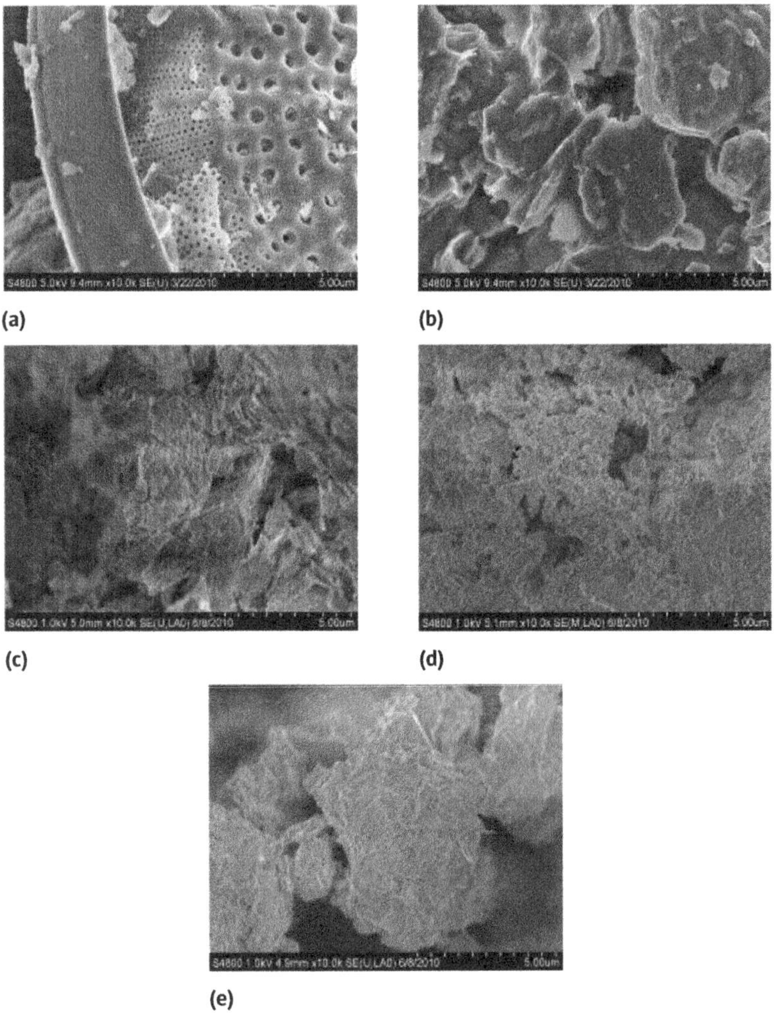

Figure 4.12 SEM images (10,000× magnification) of five insoluble substances: (a) diatomite; (b) kaolin; (c) coal powder; (d) charcoal powder; (e) insoluble particles in coastal natural pollution.

lower surface conductivity. Under saturated conditions, higher NSDD results in increased surface conductivity.

This phenomenon can be explained as follows: In the unsaturated state, moisture in the pollution layer primarily comes from the hygroscopic properties of the salt and insoluble substances. A higher ash density means more moisture is retained by the insoluble substances, leaving less available to dissolve the salt, thus reducing surface conductivity. In the saturated state, where water is abundant,

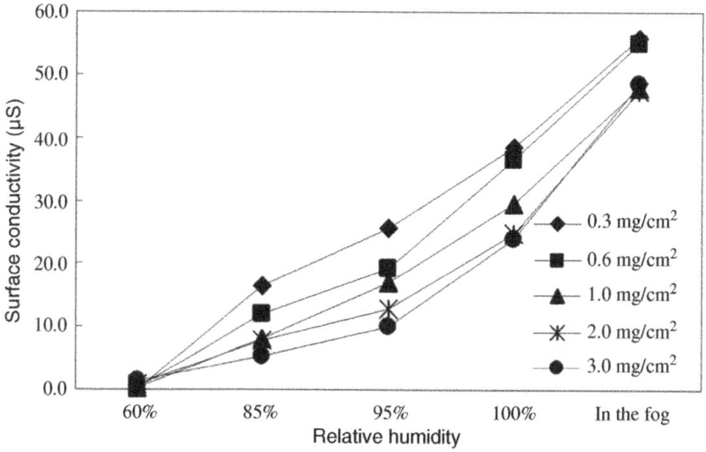

Figure 4.13 Influence of ash density on surface conductivity (25 °C, salt density: 0.05 mg/cm²).

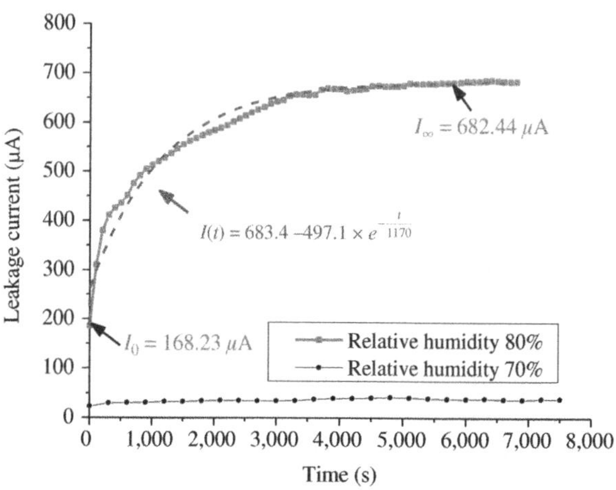

Figure 4.14 Change of leakage current at 80% and 70% relative humidity.
Source: Cao et al. (2022)/with permission of IEEE.

surface conductivity is primarily determined by the NSDD of the pollution layer at a SDD.

4.3.3 Dynamics of Water Absorption and Evaporation at Various Humidities

Figure 4.14 shows the changes of leakage currents in the specimens with 1.0 mg/cm² NSDD and a 0.1 mg/cm² ESDD at 70 and 80% relative humidity.

At the 80% relative humidity, the leakage current increased from 168.23 μA gradually to 682.44 μA. In contrast, at 70% relative humidity, the leakage current remained below 20 μA throughout the test period. According to the theoretical analysis mentioned before, the leakage current behavior during the wetting process is governed by the same dynamics as water absorption. That is, the leakage current follows an exponential growth model described by:

$$I(t) = I_{\infty} + (I_0 - I_{\infty})e^{-\frac{t}{\tau}} \tag{4.1}$$

As the applied voltage 100 V and geometry of specimens were known parameters, the time constant τ in the experiments was derived as nearly 1170 s. The measured leakage current at t = 0 s could be regarded as the initial value I_0. In addition, the equilibrium leakage current after 4500 s could be regarded as the value I_{∞} at infinite time.

$$\begin{cases} \tau = 1170\ \text{s} \\ I_{t=0} = 186.3\ \mu\text{A} \\ I_{t=\infty} = 668.2\ \mu\text{A} \end{cases} \tag{4.2}$$

The predicted leakage current curve at 80% atmospheric humidity, based on Eqs. (4.1) and (4.2), is also drawn as the dashed line in Figure 4.14. It is found that the calculated leakage current is nearly the same as the measured value. The water of the contamination layer eventually reached an equilibrium value when the velocities of water absorption and evaporation were equal.

4.4 Relationship Between Humidity, Contamination, and Leakage Current

When the insulator gets wet, a greater amount of soluble salts in the contamination layer dissolves, releasing more ions into the moisture film. This leads to increased conductivity and a higher leakage current under the same applied voltage. Under 100% relative humidity, due to the Joule heating effect of the leakage current, local dry bands are formed near the insulator end fittings, and the intermittent breakdowns of the local dry bands by the concentration of electric field will cause a sudden increase of the leakage current amplitude as well as the high-order harmonics. This phenomenon forms the basis for evaluating the degree of insulator pollution using either the leakage current amplitude or its harmonic components under saturated moisture conditions.

However, under unsaturated humidity conditions, the surface leakage current is relatively small, and no local arc occurs on the insulator surface. As a result, characteristic parameters of the leakage current defined under saturated conditions are not applicable for assessing pollution severity in these cases.

At unsaturated atmospheric humidity, the soluble salts in the contamination layer are only partially dissolved, resulting in a saturated salt solution. Assuming that the water absorption velocity of the contamination layer is denoted as $V_{\mathrm{RH}}(\mathrm{RH},\ \mathrm{ESDD},\ t)$, and the evaporation velocity as $V_{\mathrm{evop}}(t)$, the change of water content in the contamination layer m_{water} over time is:

$$\frac{dm_{\mathrm{water}}}{dt} = V_{\mathrm{RH}}(\mathrm{RH},\ \mathrm{ESDD},\ t) - V_{\mathrm{evop}}(t) \tag{4.3}$$

where t is the time.

We then proceed to analyze the water absorption and evaporation velocity of the contamination layer during the wetting process.

4.4.1 Water Absorption Velocity V_{RH}

At unsaturated atmospheric humidity and without voltage applied, the water in the contamination layers could not dissolve the soluble salt completely, so if the vapor pressure of the saturated salt solution in the contamination layer is higher than that of the atmosphere, the water content in the contamination layer will increase. A fundamental theory for studying the diffusion process is given by the Fick's diffusion theory, which is expressed as

$$J = -D \cdot \nabla c \tag{4.4}$$

where J is the diffusion flux, D is the diffusion coefficient and c is the concentration of water molecules in the contamination layer.

Considering the diffusion flux in a fixed volume element, Fick's second law is obtained:

$$\frac{\partial c}{\partial t} = \nabla \cdot (D \cdot \nabla c) \tag{4.5}$$

where t is the diffusion time.

As the wetting of contamination layers on the insulator can be regarded as a one-dimension diffusion model, the second Fick's law can be adopted as

$$\frac{\partial c}{\partial t} = D\frac{\partial^2 c}{\partial x^2} \tag{4.6}$$

Therefore, the change of water absorption of the contamination layers during wetting process is obtained by Eq. (4.6) with given boundary conditions.

The water content of contamination layer is mainly affected by soluble salt. Considering that there is a three-phase thermodynamic equilibrium system formed among soluble salt in the contamination layers, saturated salt solution, and upper atmosphere, due to both effects of ionization and hydration of the saturated salt solution, the vaporization and liquefaction of water molecules happen simultaneously, which makes the water content of the contamination layer tend to be in a dynamic equilibrium.

If moving the insulator with dry contamination layers to a test chamber with a constant atmospheric relative humidity (RH), with the increase of water absorption in the contamination layer c, the water absorption velocity ($\partial c/\partial t$) gradually decreases. However, for the energized insulators, due to the thermal effect caused by leakage current, the water content in the contamination layer cannot reach saturated, and the water absorption velocity $V_{RH}(\text{RH, ESDD}, t)$ of the pollution layer could be regarded as a constant velocity $V_{RH}(\text{RH, ESDD}, t)|_{t\approx 0}$ at the initial stage with a low water content which is independent of contamination's water absorption.

The water absorption velocity is influenced by the atmospheric humidity and the insulator pollution degree. With a higher pollution degree, more soluble salt particles exist in the contamination layer, which results in a higher diffusion coefficient D and higher water absorption velocity. On the other hand, the higher atmospheric humidity in the atmosphere is, the higher the concentration of water molecules in the air, which also results in a higher water absorption velocity. The water absorption velocity of the contamination layer could be expressed as,

$$V_{RH}(\text{RH, ESDD}) = \text{ESDD} \times k_{RH} \times (\text{RH} - \text{RH}_0) \tag{4.7}$$

where RH is the atmospheric relative humidity and k_{RH} is the coefficient. Assuming that the atmospheric RH above the saturated salt solution is RH_0 at the equilibrium state, without voltage applied, when the atmospheric RH is higher than RH_0, the contamination layer absorbs water, and when the atmospheric RH is below RH_0, the water in the contamination layer will evaporate.

4.4.2 Evaporation Velocity, V_{evop}

The evaporation velocity per unit surface can be expressed as:

$$V_{evop} = \frac{P}{\Delta H} \tag{4.8}$$

where ΔH is the heat of evaporation of water (mg/cm^3), P is the heating power per unit area of the salt solution layer (W/cm^2) and, with voltage applied, it can be expressed as:

$$P = \frac{UI}{S} = \frac{KU^2}{fS} \tag{4.9}$$

where U and I are the applied voltage (kV) and leakage current (mA) respectively; S is the surface area of the insulator (cm^2); K is the surface conductivity of the contamination layers (μS); and *f* is the shape factor of the insulator, which can be described as:

$$f = \int_0^L \frac{dx}{\pi D(x)} \tag{4.10}$$

where *L* is the creepage distance of the insulator (cm), and $D(x)$ refers to the diameter (cm) where the creepage distance is *x*.

The relationship between the surface conductivity *K* and water content $m_{\text{water}}(\text{mg/cm}^2)$ has been studied in Chapter 3, as shown in Figure 4.15. Without rain or fog partials, the water content in the contamination does not reach its maximum value (dashed line in Figure 4.15), the surface conductivity is proportional to the water content in the contamination layer and the surface conductivity could be expressed by Eq. (4.11):

$$K(m_{\text{water}}) = \frac{\sigma_{\text{salt}-s}}{\rho_{\text{salt}-s}} m_{\text{water}} = k_1 m_{\text{water}} \tag{4.11}$$

where $\sigma_{\text{salt}-s}$ and $\rho_{\text{salt}-s}$ are the conductivity (mS/cm) and density (mg/cm^3) of saturated NaCl solution, and k_1 is the relationship coefficient, 210.9 $\mu\text{S cm}^2/\text{mg}$.

Therefore, combined Eqs. (4.8)–(4.11), the evaporation velocity per unit surface can be expressed as,

$$V_{\text{evop}} = \frac{P}{\Delta H} = \frac{k_1 m_{\text{water}} U^2}{\Delta H \times f \times S} \tag{4.12}$$

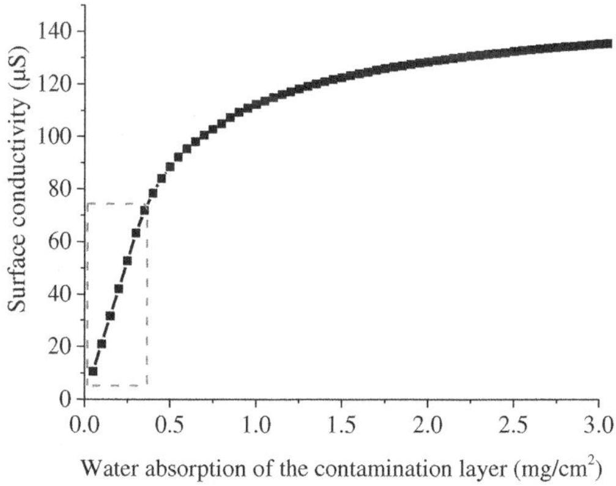

Figure 4.15 The relationship between the surface conductivity and water content in the contamination layer (ESDD = 0.1 mg/cm²). *Source*: Cao et al. (2022)/with permission of IEEE.

4.4.3 Changing of Water Absorption in the Contamination Layer at Different Atmospheric Humidity

As the water absorption velocity is $V_{RH}(\mathrm{RH},\ \mathrm{ESDD})$ and evaporation velocity V_{evop} has been obtained by Eqs. (4.7) and (4.12), then substitute them into Eq. (4.3), the change in water absorption in the contamination layer could be expressed as:

$$\frac{dm_{water}}{dt} = V_{RH}(\mathrm{RH},\ \mathrm{ESDD}) - \frac{k_1 m_{water} U^2}{\Delta H \times f \times S} \tag{4.13}$$

If the time constant τ is defined as:

$$\tau = \frac{\Delta H \times S \times f}{k_1 U^2} \tag{4.14}$$

then, Eq. (4.13) could be written as:

$$\frac{dm_{water}}{dt} = V_{RH}(\mathrm{RH},\ \mathrm{ESDD}) - \frac{m_{water}}{\tau} \tag{4.15}$$

Equation (4.15) is a first-order partial differential equation, and its general solution is:

$$m_{water}(t) = V_{RH}(\mathrm{RH},\ \mathrm{ESDD}) \times \tau \times \left(1 + C_1 e^{-\frac{t}{\tau}}\right) \tag{4.16}$$

where C_1 is the undetermined coefficient.

According to Eq. (4.16), if the atmospheric RH is RH_1 and RH_2, respectively, when the wetting reaches an equilibrium level ($t = \infty$), and the equilibrium water in the contamination layer at RH_1 and RH_2 is $\tau_{V_{RH}(\mathrm{RH}_1,\ \mathrm{ESDD})}$ and $\tau_{V_{RH}(\mathrm{RH}_2,\ \mathrm{ESDD})}$, respectively.

If changing the atmospheric relative humidity from RH_1 to RH_2 suddenly, the change of water content from the initial equilibrium value m_0 to the equilibrium value m∞ with the following boundary conditions:

$$\begin{cases} m_0 = \tau k_{RH} \mathrm{ESDD}(\mathrm{RH}_1 - \mathrm{RH}_0) & t = 0 \\ m_\infty = \tau k_{RH} \mathrm{ESDD}(\mathrm{RH}_2 - \mathrm{RH}_0) & t = \infty \end{cases} \tag{4.17}$$

According to the first-order differential Eq. (4.15) and boundary conditions (4.17), the change in water content during the wetting process could be obtained and the change of water content in the contamination layers obeyed an exponential approach function, as shown in Figure 4.16 and Eq. (4.18).

$$m_{water}(t) = m_\infty + (m_0 - m_\infty)e^{-\frac{t}{\tau}} \tag{4.18}$$

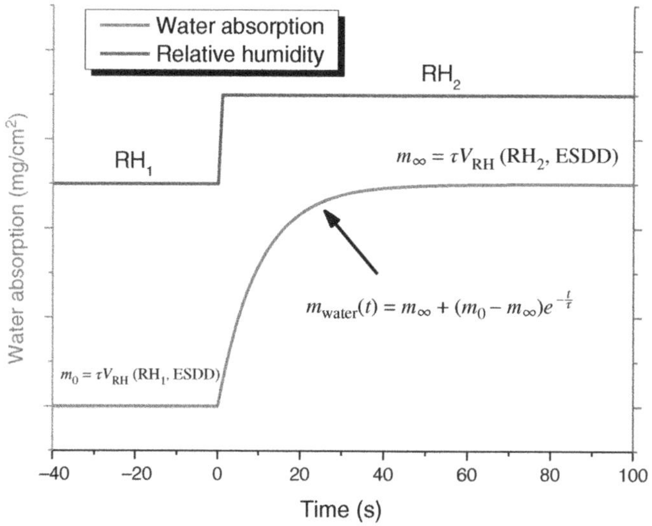

Figure 4.16 Water absorption of the contamination layer.

4.4.4 Relationship Between Leakage Current, Insulator Contamination, and Atmospheric Humidity

As discussed before, with voltage applied, the change of water content in the contamination layers obeyed an exponential approach function. Then, combined with the applied voltage, the leakage current I could be obtained from the water content in the contamination layer m_{water}:

$$I = \frac{K(m_{water})U}{f} = \frac{k_1 m_{water} U}{f} \tag{4.19}$$

If the atmospheric humidity increases from RH_a to RH_b, the water content in the contamination layer gradually increases exponentially to an equilibrium value at which the evaporation velocity caused by the leakage current is the same as the water absorption velocity at RH_b. When the atmospheric humidity decreases from RH_b to RH_a, the leakage current gradually decreases back to the equilibrium value at RH_a, as shown in Figure 4.17. With same pollution degree and atmospheric humidity, the equilibrium leakage current at RH_a is the same.

As the change of water content in the contamination layers obeyed an exponential approach function, the time required for the leakage current from 0 to the equilibrium value is 3~5 τ. Considering that the voltage applied on the operating insulator is high (unified specific creepage distances are 22.0~53.7 mm/kV),

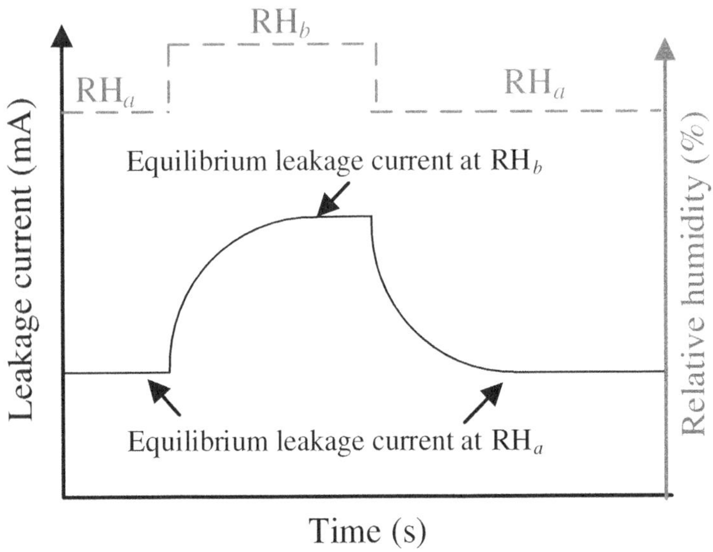

Figure 4.17 The change of leakage current with humidity. *Source*: Cao et al. (2022)/with permission of IEEE.

6.0~14.5 kV will be applied on a single cap and pin insulator with 320 mm leakage distance, and the constant time τ is several seconds. Hence, for live monitoring of in-service insulators, the transient process can be ignored, and the measured leakage current can be regarded as its equilibrium value under current humidity conditions.

Thus, the real-time leakage current $I(\text{RH})$ can be modeled as a function of ambient humidity RH and the insulator ESDD:

$$I(\text{RH}) = p \times (\text{RH} - \text{RH}_0) \tag{4.20}$$

where p and RH_0 are the fitted coefficients. The parameter p is the slope between leakage current and humidity, which could also be calculated by:

$$p = \frac{\Delta H \times S \times f \times k_{\text{RH}}}{U} \times \text{ESDD} \tag{4.21}$$

Since the slope p is mainly affected by the pollution degree of the insulator, it could be used to monitor the insulator. The hygroscopicity of different pollution components might be different, which may result in different $k_{\text{RH-values}}$ under the same pollution degree. Fortunately, in the specific area with the same typical pollution source, the proportion of insulator contamination component does not change with time.

4.5 Summary

In this chapter, two experimental systems were designed to investigate insulator wetting characteristics: a pollution test system with controllable temperature and humidity, based on the mixing flow method using a humidity generator, and a small-scale environmental system with controlled temperature, humidity and pressure employing the equilibrium vapor pressure method.

The former system allows control of RH from the ambient level up to 99%, with humidity fluctuations within ±3%. The temperature range extends from the ambient temperature to ambient temperature +30 °C (but not exceeding 60 °C), with a temperature fluctuation within ±1 °C.

The latter system, whose humidity is governed by saturated salt solutions rather than ambient conditions, provides stable humidity levels that are not continuously adjustable but fluctuate within ±1%. Its temperature can be continuously controlled within a range of 0 °C to 100 °C, also maintaining a fluctuation within ±1 °C.

The investigation into the influencing factors of insulator wetting characteristics was conducted using both systems, with surface conductivity of the insulator employed as a key physical parameter for analysis.

Bibliography

Cao, B., and Wang, L. Wetting characteristics of artificial contamination on the hydrophobic surface, Eng. Fail. Anal., 2019, 100, 428–438.

Cao, B., Yin, F.H., Yang, D.M. et al. Characterisation parameter of atmospheric pollutant concentration for external insulation, IET Gener. Transm. Distrib., 2020, 14 (20), 4547–4552.

Cao, B., Shen, S., Li, Z. et al. Insulator contamination monitoring based on its hygroscopicity under unsaturated humidity, IEEE Trans. Instrum. Meas., 2022, 71, 1–8.

Chisholm, W.A., Buchan, P.G., and Jarv, T. Accurate measurement of low insulator contamination levels (in English), IEEE Trans. Power Del., 1994, 9 (3), 1552–1557.

He, H., Luo, D., Lee, W.-J. et al. A contactless insulator contamination levels detecting method based on infrared images features and RBFNN, IEEE Trans. Ind. Appl., 2019, 55 (3), 2455–2463.

He, Z., Gao, F., Tu, Z. et al. Analysis of natural contamination components and sources of insulators on ±800 kV DC lines, Electr. Power Syst. Res., 2019, 167, 192–198.

Ibrahim, M.E. and Abd-Elhady, A.M. Rogowski coil transducer-based condition monitoring of high voltage insulators, IEEE Sens. J., 2020, 20 (22), 13694–13703.

IEC TS 60815-1, Selection and Dimensioning of High-voltage Insulators Intended for Use in Polluted Conditions - Part 1: Definitions, Information and General Principles, 2008.

Kokkinaki, O., Klini, A., Polychronaki, M. et al. Assessing the type and quality of high voltage composite outdoor insulators by remote laser-induced breakdown spectroscopy analysis: a feasibility study, Spectrochim. Acta B At. Spectrosc., 2020, 165, 105768.

Lan, L., Mu, L., Yuan, X. et al. Studies on metallurgical contamination accumulation characteristics on ceramic insulator of 500 kV AC transmission line, IET Sci. Meas. Technol., 2019, 13 (5), 722–728.

Li, Z., Yin, F., Cao, B. et al. Pollution flashover performance of RTV coatings with partial damage, Int. J. Electr. Power Energy Syst., 2020, 121, 106102.

Liu, L., Guo, C., Tu, Y. et al. Differential evolution fitting-based optical step phase thermography for micron thickness measurement of atmospheric corrosion layer, IEEE Trans. Ind. Inform., https://doi.org/10.1109/TII.2019.2955493.

Montoya, G., Ramirez, I., and Montoya, J.I. Correlation among ESDD, NSDD and leakage current in distribution insulators, IEE Proc. Gener. Transm. Distrib., 2004, 151 (3), 334–340.

Mussina, D., Irmanova, A., Jamwal, P.K. et al. Multi-modal data fusion using deep neural network for condition monitoring of high voltage insulator, IEEE Access, 2020, 8, 184486–184496.

Palangar, M.F., Amin, U., Bakhshayesh, H. et al. Identification of composite insulator criticality based on a new leakage current diagnostic index, IEEE Trans. Instrum. Meas., 2021, 70, 1–10.

Qin, X., Zhang, F., Chen, S. et al. Characterization of hygroscopic insulator contamination via laser-induced breakdown spectroscopy, IEEE Trans. Plasma Sci., 2021, 49 (3), 1166–1172.

Salem, A.A., Abd-Rahman, R., Al-Gailani, S.A. et al. Risk assessment of polluted glass insulator using leakage current index under different operating conditions, IEEE Access, 2020, 8, 175827–175839.

Sit, K., Das, A.K., Mukherjee, D. et al. Condition monitoring of overhead polymeric insulators employing hyperbolic window stockwell transform of surface leakage current signals, IEEE Sens. J., 2021, 21 (9), 10957–10964.

Suda, T. Frequency characteristics of leakage current waveforms of a string of suspension insulators, IEEE Trans. Power Del., 2005, 20 (1), 481–487.

Wu, J., Sfarra, S., and Yao, Y. Sparse principal component thermography for subsurface defect detection in composite products, IEEE Trans. Ind. Inform., 2018, 14 (12), 5594–5600.

Yang, D., Cao, B., Li, Z. et al. On-line monitoring, data analysis for electrolytic corrosion of ±800 kV high voltage direct current insulators, Int. J. Electr. Power Energy Syst., 2021, 131, 107097.

Yin, C., Guo, Y., Zhang, X. et al. A novel method for visualizing the pollution distribution of insulators, IEEE Trans. Instrum. Meas., 2021, 70, 1–8.

Zhang, Y., Huang, X., Jia, J. et al. Detection and condition assessment of icicle bridging for suspension glass insulator by image analysis, IEEE Trans. Instrum. Meas., 2020, 69 (10), 7458–7471.

5

Temperature Differences and Their Effect on Wetting Behavior

5.1 Introduction

In addition to the influence of ambient humidity, as analyzed in Chapters 2 and 4 of this book, temperature also plays a significant role in the wetting behavior of pollution layers. In particular, variations in ambient temperature can cause the surface temperature of insulators to fall below the surrounding air temperature, leading to condensation on the insulator surface. Such condensation-induced wetting is also frequently observed in arid and semiarid regions, where ambient humidity is generally low and insufficient by itself to cause surface wetting. In these regions, pronounced day–night temperature differences can result in condensation that moistens the pollution layer on insulators. This unexpected wetting behavior can substantially alter surface conductivity and thereby affect the insulation performance. On the other hand, for leakage current–based monitoring systems used to assess the performance of external insulation, understanding the impact of temperature differences on wetting behavior is crucial for improving both insulator flashover resistance and the reliability of early warning systems for pollution flashover. To this end, this chapter investigates the effect of temperature-induced condensation on water absorption in contaminated layers.

In this chapter, laboratory experiments were conducted in a controlled artificial climate chamber, simulating various temperature difference scenarios under moderate humidity. The results demonstrate a strong positive correlation between the initial temperature difference and the amount of water absorbed by the pollution layer. Furthermore, by analyzing leakage current trends and estimating water absorption over time, a modified contamination evaluation model is proposed. This model integrates environmental conditions and surface interface behavior, enhancing the accuracy of online contamination monitoring—especially in inland regions characterized by low humidity and large diurnal temperature fluctuations.

5.2 Experimental Arrangement

5.2.1 Evaluation Method of Wetting Condition Based on Temperature Difference

For the hydrophilic insulating materials such as glass and porcelain, the lack of hydrophobicity leads to the formation of continuous contamination film during the wetting process, as shown in Figure 5.1. The more soluble salts on the surface of insulators are dissolved during the wetting process, the more ions will be introduced into the contamination solution, resulting in higher surface conductivity. For insulators in service, the increase of surface conductivity can also result in the increase of leakage current, which in turn promotes the evaporation of the contamination solution. Therefore, the wetting process of insulators is a dynamic in nature.

Under conditions of moderate humidity, collision-based wetting does not occur because the water vapor in the air has not yet reached saturation. In this case, wetting of the contamination layer is primarily driven by temperature differences and water absorption. Moreover, the primary mechanism of moisture loss in the contamination layer is evaporation caused by the thermal effects of leakage current. Therefore, when an insulator is exposed to an environment where the ambient temperature is higher than that of the insulator surface, a typical leakage current profile under the combined influence of temperature difference, moisture, and evaporation is observed, as shown in Figure 5.2.

The leakage current under the condition of temperature difference can be divided into three phases:

1. Phase A–B: When the temperature difference between the insulator surface and the ambient environment is large, the wetting rate of the contamination layer—driven by the temperature gradient—exceeds the evaporation rate caused by the thermal effect of leakage current. As a result, the leakage current gradually increases, as shown in section A–B in Figure 5.2.

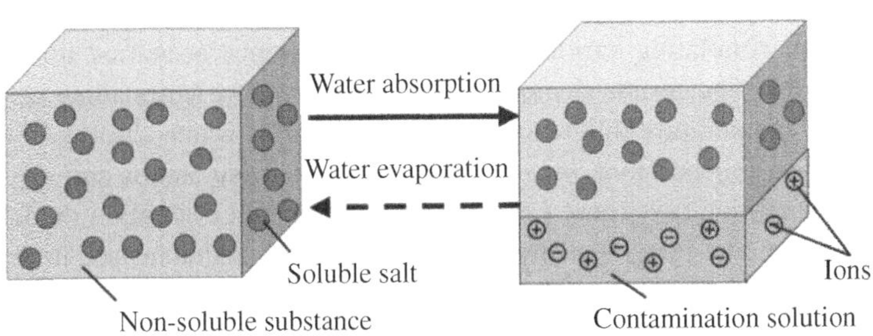

Figure 5.1 Wetting process of contamination layers on insulators. *Source*: Wang et al. (2017)/with permission of IEEE.

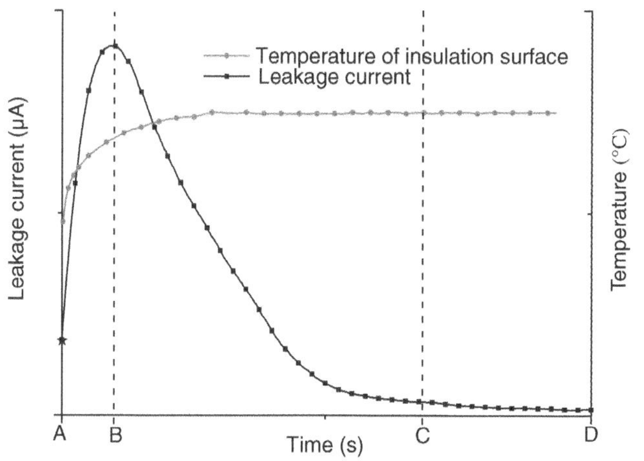

Figure 5.2 Typical leakage current under a temperature gradient. *Source*: Wang et al. (2017)/with permission of IEEE.

2. Phase B–C: When the evaporation rate surpasses the wetting rate of induced by the temperature difference, the leakage current begins to decline. Simultaneously, the temperature difference between the insulator surface and the ambient air decreases, as shown in section B–C in Figure 5.2.
3. Phase C–D: When the insulator surface temperature equals the ambient temperature and the wetting rate balances the evaporation rate, the leakage current will reach a dynamic equilibrium, as shown in section C–D in Figure 5.2.

Therefore, for a given insulator, a larger temperature difference results in greater water absorption by the contamination layer, leading to higher surface conductivity and an increase in leakage current. Moreover, if the soluble salt in the contamination is dissolved completely, the leakage current can be utilized to characterize the contamination degree of the insulation surface.

As shown in Figure 5.3, the moisture content under a temperature gradient consists of two independent parts. One is the condensation caused by the temperature difference, which gradually decreases per unit time as the temperature difference narrows. The other is the adsorbed moisture due to ambient humidity, which occurs at a constant rate regardless of time. Therefore, the total water absorption can be expressed as:

$$m_{\text{total}} = m_{\text{temp}} + m_{\text{humi}} \tag{5.1}$$

where m_{temp} refers to the amount of water absorbed caused by the temperature difference, and m_{humi} denotes the amount of water absorbed due to ambient humidity.

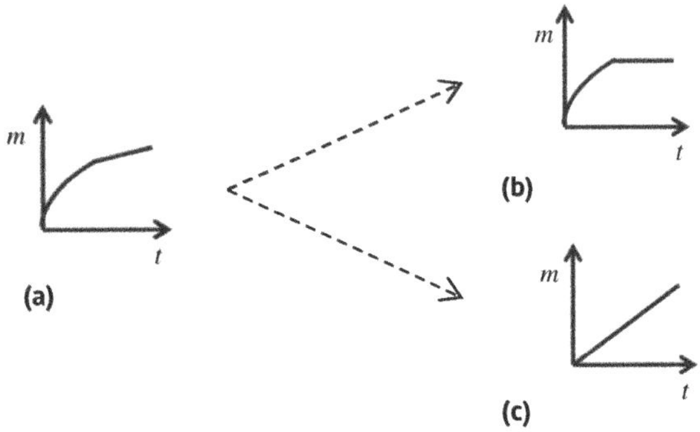

Figure 5.3 Water absorption of contamination layers under a temperature difference: (a) the total amount of water absorbed; (b) the condensation caused by temperature difference; (c) the absorbed moisture due to ambient humidity. (*Source*: Wang et al. (2017)/with permission of IEEE). *Note:* The figure is plotted with time on the horizontal axis and the amount of water absorption on the vertical axis.

5.2.1.1 Calculation of the Total Water Absorption Caused by Temperature Difference

During the entire period when the temperature difference takes effect, the water absorption of the contamination can be divided into two independent components: one is the evaporated portion of the contamination solution due to the thermal effect of the leakage current, and the other is the residual moisture remaining in the contamination layer. Therefore, the total amount of water caused by temperature difference can be expressed as below:

$$m_{\text{total}} = m_{\text{evap}} + m_{\text{resi}} \tag{5.2}$$

where m_{evap} signifies the evaporated portion of the contamination solution while m_{resi} refers to the residual moisture remaining in the contamination.

Correspondingly, the water amount caused by temperature difference is as follows:

$$m_{\text{temp}} = m_{\text{evap}} - m_{\text{humi}} + m_{\text{resi}} \tag{5.3}$$

5.2.1.2 Calculation of the Residual Water in the Contamination Layer

Although the actual amount of water on a contaminated surface can be measured using a balance, the water quantity on the insulation surface in moderate humidity conditions is very small. If the mass of residual water in the contamination is measured by the balance, the measurement error will be significant.

In artificial contamination samples, water absorbed by the contamination layer dissolves the soluble salts. If the salts are not fully dissolved, the contamination solution reaches saturation, and the water absorption becomes directly

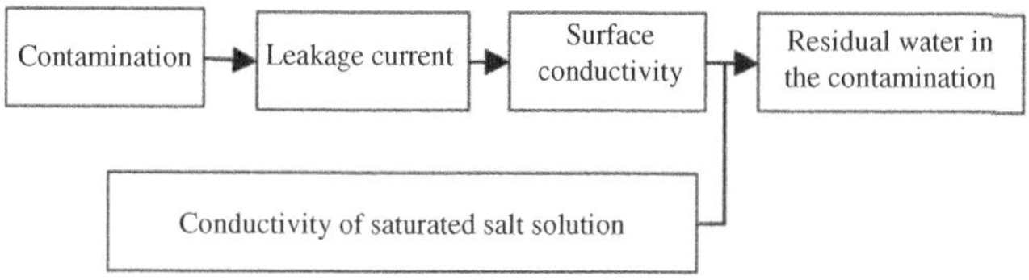

Figure 5.4 The calculation flow chart of the residual water in the contamination layer. *Source:* Wang et al. (2017)/with permission of IEEE.

related to the surface conductivity. Since surface conductivity can be measured precisely, the residual water in the contamination layer can be estimated based on the relationship between water absorption and surface conductivity with the conversion method illustrated in Figure 5.4.

The residual water in the contamination layer can be calculated by:

$$m_{\text{resi}} = \rho_{\text{water}} \times S \times h \tag{5.4}$$

where ρ_{water} refers to the density of water; S means the area 1 cm^2 and h denotes the thickness of the water film after damp.

Moreover, if the contamination solution is not saturated, the relationship between the surface conductivity and the conductivity of the saturated solution can be expressed as:

$$K = \sigma \times h \tag{5.5}$$

where K signifies the surface conductivity and σ refers to the conductivity of the saturated solution. Therefore, the relationship between surface conductivity and water absorption can be established, as shown in Figure 5.5.

The surface conductivity K can be obtained by the leakage current using the following equation:

$$K = \frac{I}{U} f = Gf \tag{5.6}$$

where I is the leakage current, U is the voltage applied on the insulators, and f denotes the shape factor of insulator.

A temperature correction can be applied to the measured surface conductivity using the following formula:

$$K_{20} = k_t \cdot K_t \tag{5.7}$$

where K_{20} refers to the surface conductivity corrected under 20 °C, K_t means the surface conductivity measured at temperature t, and k_t signifies the temperature coefficient.

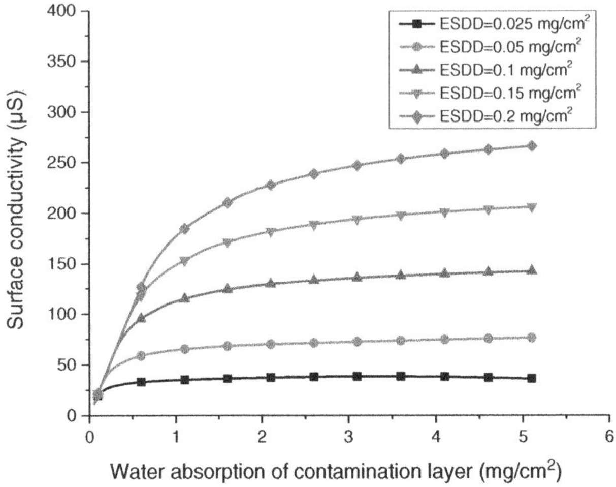

Figure 5.5 Relationship between surface conductivity and water absorption of the contamination layer. *Source*: Wang et al. (2017)/with permission of IEEE.

5.2.1.3 Calculation of Surface Evaporation

When the insulator gets wet, the soluble salts may partially or completely dissolve, which increases the surface conductivity and consequently the leakage current. This results in the evaporation of some surface water.

The mass of evaporated water can be calculated by:

$$m_{\text{evap}} = \int_0^T \frac{P}{k_1} dt \tag{5.8}$$

where k_1 is the latent heat of vaporization at the given temperature, ranging from 2441.7 to 2465.6 J/g, and P refers to the heating power of the contamination surface per unit area caused by the leakage current. The heating power P can be expressed as:

$$P = E \times J \tag{5.9}$$

where E refers to the electric field strength and J denotes the surface current density.

5.2.1.4 Calculation of the Hygroscopic Behavior

The contamination layer exhibits hygroscopic behavior even under relatively low humidity, especially for insulators in service. When the leakage current reaches equilibrium, a balance is established between the evaporation due to thermal effects and the hygroscopic water uptake from the air. At this equilibrium, the amount of absorbed moisture can be calculated.

5.2.2 Test Specimens, Facilities, and Procedure

5.2.2.1 Fog Chamber and Humidity Control System

The dampening process of the insulator is carried out inside a compact fog chamber with controllable temperature and humidity, as shown in Figure 5.6. A heating wire is arranged along the chamber walls to ensure uniform air heating. Several pipelines connected to an ultrasonic humidifier are installed around the fog chamber and the inner circulation fan is employed to achieve uniform humidification. Moreover, the control of temperature and humidity can be achieved by the control system.

To generate a temperature difference, the test sample is first placed in a chamber with constant temperature and humidity for a designated period, allowing its temperature to drop below the ambient level. The adjustable temperature range of this chamber is −45 to 160 °C and the relative humidity range is 0–100%.

5.2.2.2 Test Sample Preparation

The test sample used is a glass plate with dimensions of 36 × 20 × 0.8 cm. Electrodes made of aluminum foil, 1 and 2 cm in width, are affixed to both ends. The effective polluted area between the electrodes measures 33 × 20 cm.

5.2.2.3 Applied Voltage

For the insulators in actual operation, the wetting process is a dynamic balance between water absorption and evaporation. Since the evaporation rate depends

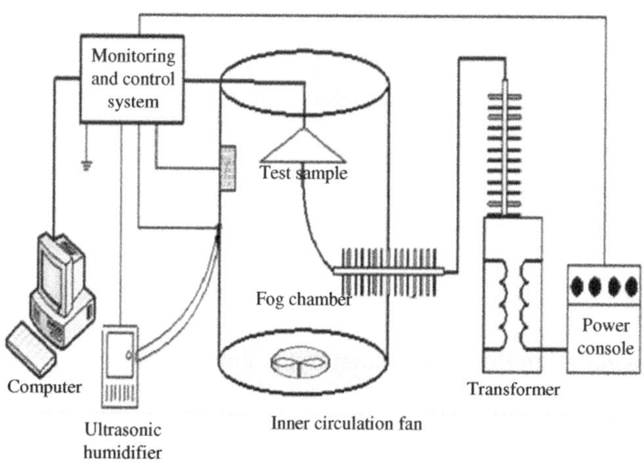

Figure 5.6 Structure diagram of testing system.

on the leakage current, a proper test voltage must be selected to reflect realistic conditions. NaCl dissolves easily in water, and once the surface becomes wet, conductivity rises rapidly. If the applied voltage is too high, it may lead to partial discharge and increased thermal evaporation, which would deviate from natural wetting- evaporation behavior.

To better simulate realistic operating conditions—where the leakage current rarely exceeds 10 mA without partial discharge—a voltage of 100 V is applied in this test. Under this voltage, the leakage current of the test sample is limited to below 5 mA.

5.2.2.4 Test Procedure

The test procedure is as follows:

1. Preparation of contamination layer: Uniform contamination is applied to the test piece. The Equivalent Salt Deposit Density is 0.1 mg/cm^2, and the Non-soluble Deposit Density is 1.0 mg/cm^2. The contamination consists of NaCl as the soluble component and kaolin as the insoluble component.
2. Temperature preconditioning: Place the test piece into the box with constant temperature and 50% relative humidity to maintain the contamination layer dry. The temperature is set according to the test requirements. As heat conduction through the glass requires time, the sample is preconditioned for at least 40 minutes to ensure uniform cooling.
3. Setting ambient conditions: Set and stabilize the temperature and the humidity in the fog chamber according to test requirements.
4. Test execution: Put the glass plate into the fog chamber quickly and then add voltage between two electrodes of the test piece.
5. Data acquisition: Continuously measure and record leakage current and temperature of the test piece during the test period and the collection interval is set to 1 second.

The initial temperature differences between the test sample and ambient environment are set to 15, 10, 5, and 0 °C respectively. The ambient temperature maintains at 25 °C.

5.3 Results

5.3.1 Temperature Variation Between Insulator and Environment

When there is a temperature difference between the insulator surface and the ambient temperature, intense energy exchange occurs between the insulator surface and the ambient air. Consequently, the temperature of the insulator surface will gradually rises to approach the ambient temperature.

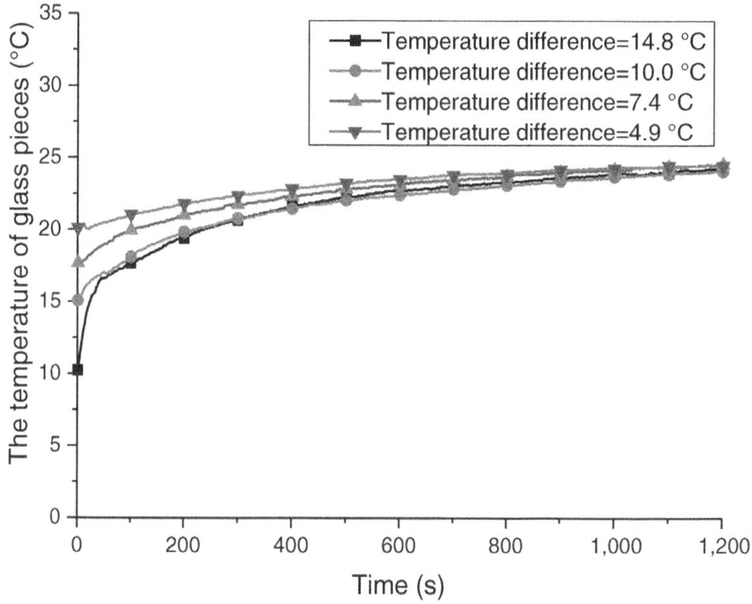

Figure 5.7 The changes of the plate glass temperature under different initial temperature differences. *Source*: Wang et al. (2017)/with permission of IEEE.

The temperatures of the glass plate in the center and the four corner positions were measured by the contact thermocouple thermometer, and the average value was taken as the glass plate temperature. The changes of the temperatures of glass plate with different initial temperatures and time period are shown in Figure 5.7. If the temperature curve of the glass plate with initial 15 °C temperature difference is set as the base curve, other curves can be shifted forward along the time axis to align the initial temperature with the base curve, as shown in Figure 5.8. It can be found that the curves are basically coincident, indicating similar temperature variation trends.

The experimental results show that in a certain temperature range, the temperature change rate of the glass plate decreases with the decrease of temperature difference. When the initial temperature is low, the heat exchange process between the glass plate and environments is more intense initially, but subsequent temperature changes are similar.

5.3.2 Effect of Temperature Difference on Water Absorption

Under the condition of 70% relative humidity and 25 °C room temperature, four experiments with different initial temperature differences were conducted. The leakage current is shown in Figure 5.9. It can be concluded that under

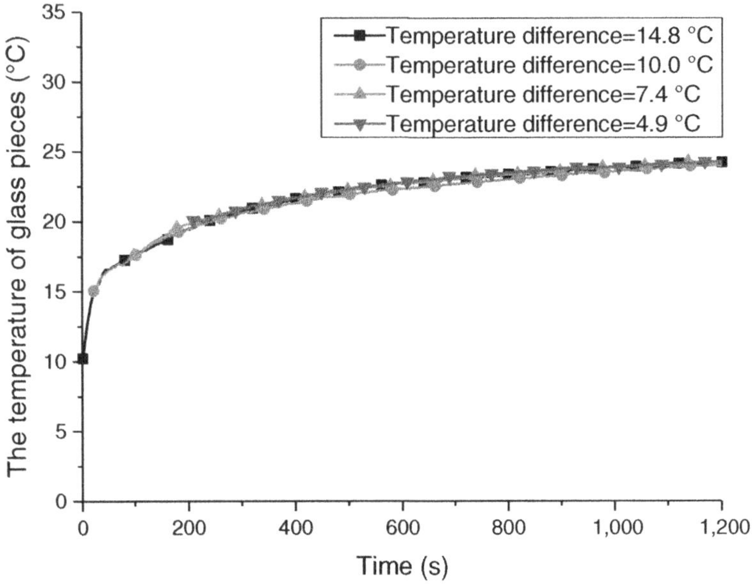

Figure 5.8 The changes of the plate glass temperature under different original temperature differences after time shift. *Source*: Wang et al. (2017)/with permission of IEEE.

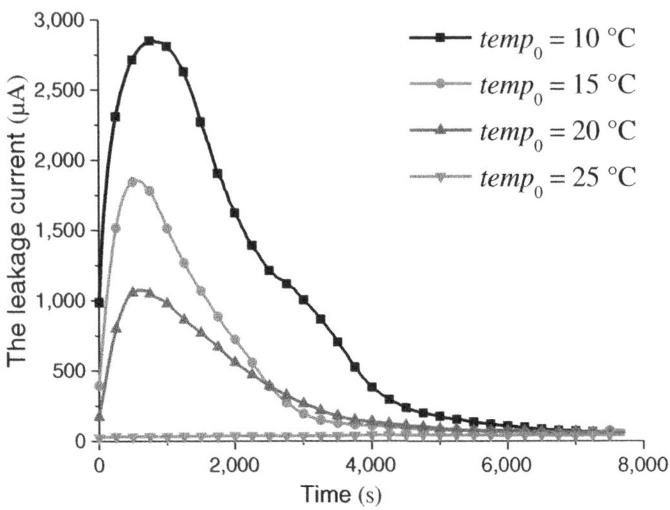

Figure 5.9 The changes of the leakage current under different initial temperature differences. *Source*: Wang et al. (2017)/with permission of IEEE.

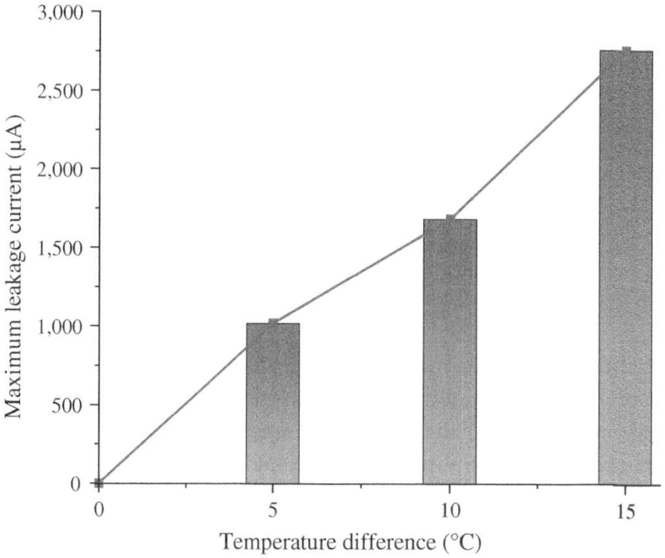

Figure 5.10 Maximum of leakage current under different temperature differences.

different temperature differences, the greater the temperature difference is, the larger the leakage current will be. When there is no temperature difference, the test sample under 70% relative humidity conditions is difficult to get damp.

The maximum leakage current is proportional to the temperature differences, as shown in Figure 5.10. The larger the temperature differences are, the larger the maximum value of leakage current will be. However, the leakage current of insulator surface depends not only on the pollution degree, but also on the water content of the contamination layer. Therefore, leakage current alone is not a sufficient parameter to characterize pollution degree, necessitating further analysis of temperature difference effects on water absorption.

5.3.3 Quantitative Effect of Temperature Difference on Water Absorption of Pollution Layer

The amount of condensation caused by temperature difference under the condition of 70% relative humidity and 25 °C room temperature is calculated from the leakage current, as shown in Figure 5.11. When the test piece is initially at a low temperature than the environment, the contamination layer gradually absorbs moisture over time. The rate of condensation decreases as the rate of the temperature change decreases. The experimental results show that the water absorption of the contamination layer is related to the temperature difference and even the

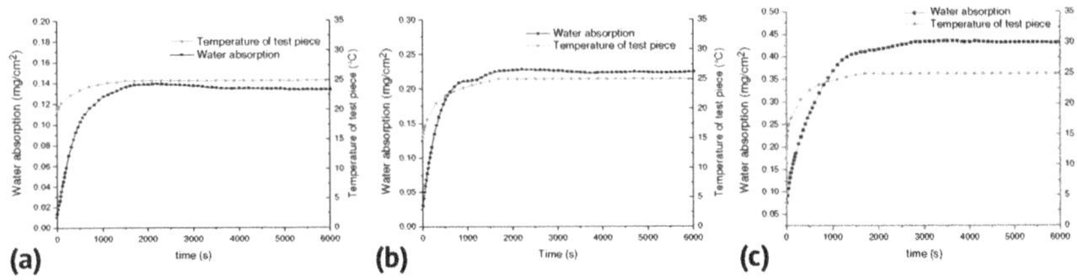

Figure 5.11 Water absorption of pollution layer over time under initial temperature differences: (a) the temperature difference is 5 °C; (b) the temperature difference is 10 °C; (c) the temperature difference is 15 °C. *Source*: Wang et al. (2017)/with permission of IEEE.

small temperature difference between the surface of the insulator and ambient temperature will cause the increase of the water absorption of the polluted layers.

Previous findings indicate that if the insulator temperature is lower than the dew point temperature of the environment, the insulators will reach saturation moisture more quickly. Dew point is a meteorological concept that refers to the temperature at which the gaseous water in the air is saturated and condensed into liquid water at a constant pressure. The dew point temperature is calculated as:

$$\begin{cases} H = (\lg H_r - 2)/0.4343 + 17.62\theta_r/(243.12 + \theta_r) \\ \theta_{DP} = 243.12H/(17.62 - H) \end{cases} \tag{5.10}$$

where H_r is relative humidity, θ_r is environment temperature, and θ_{DP} is the dew point temperature.

In the study, with 70% humidity and 25 °C, the dew point temperature is 19.1 °C. Results indicate that water absorption does not strongly correlate with dew point. Even if the temperature is below dew point, the contamination layer might not be saturated, which means that the pollution is not completely dissolved, but the presence of temperature difference increases water absorption.

The effect of temperature difference on the wetting of the pollution layers can be quantitatively analyzed based on the relationship between the water absorption and the temperature difference, which under 70% relative humidity conditions are shown in Figure 5.12. It can be concluded that the water absorption of the pollution layer is proportional to the temperature difference.

5.3.4 Relationship Between Water Absorption Rate of Contamination Layers and the Humidity at Interface

To analyze the relationship between the water absorption rate of contamination layers and the humidity at the interface, the interface humidity RH_{inter} is first calculated by:

$$RH_{inter} = \frac{\rho_w}{\rho_{w-max}} \times 100\% \tag{5.11}$$

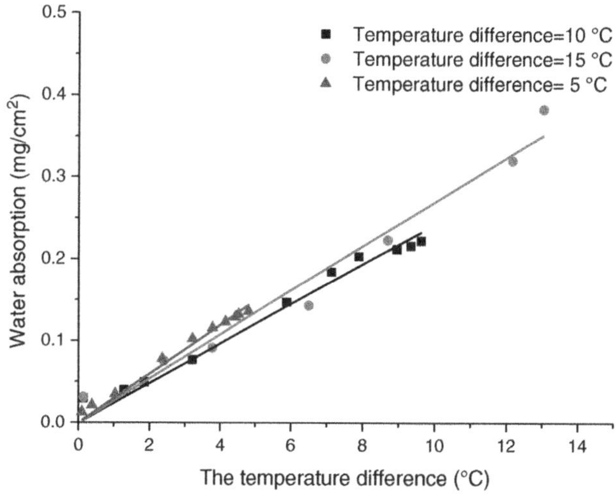

Figure 5.12 Changes of water absorption over temperature difference. *Source:* Wang et al. (2017)/with permission of IEEE.

where ρ_w and $\rho_{w-\max}$ represent absolute humidity and saturated absolute humidity respectively. $\rho_{w-\max}$ is calculated by,

$$\rho_{w-\max} = \frac{M}{RT(\text{temp})} p_s(\text{temp}) \tag{5.12}$$

where M is the molar mass of water (18 g/mol), R is gas constant (0.0623665 mmHg m^3/°C°C /mol), temp and T(temp) are the atmospheric temperature in Celsius and in Kelvin respectively, and p_s(temp) is the saturation vapor pressure. T(temp) and p_s(temp) could be calculated by:

$$\begin{cases} T(\text{temp}) = 273.15 + \text{temp} \\ p_s(\text{temp}) = (0.61078 \times 7.501) e^{\frac{17.2694\text{temp}}{238.3+\text{temp}}} \end{cases} \tag{5.13}$$

If the insulator temperature is lower than the ambient dew point, water vapor saturates and excess water condenses, causing calculated humidity RH_{inter} to exceed 100%. Considering the degree of water vapor saturation affects the moisture rate of the contamination layer, and the higher the temperature difference is, the faster the moisture rate will be. This "apparent humidity" RH_{inter} represents water molecules at the hypothetical interface and reflects the magnitude of the temperature difference.

The water absorption rate v_{water} over time interval Δ_t is:

$$v_{\text{water}} = \frac{m_{t+\Delta t} - m_t}{\Delta t} \tag{5.14}$$

where $m_{t+\Delta t}$ and m_t is the accumulated water absorption at time $t + \Delta t$ and t.

The moisture rate v_{water} is also calculated and the change of moisture rate with different initial temperature difference is shown in Figure 5.13. Regardless of the initial temperature differences, with the decrease of the temperature difference, both the apparent humidity RH_{inter} and the moisture rate of the contamination layer decreased.

To quantitatively reveal the relationship between the apparent humidity at the interfaces and the water absorption rate of the contamination layer, statistical analysis was carried out based on 4,800 sets of experimental data. As shown in Figure 5.14, the moisture rate of the contamination layer v_{water} increases with the apparent humidity RH_{inter} linearly, and the relationship could be fitted by,

$$v_{water} = 0.0072 \times (RH_{inter} - 72.6) \tag{5.15}$$

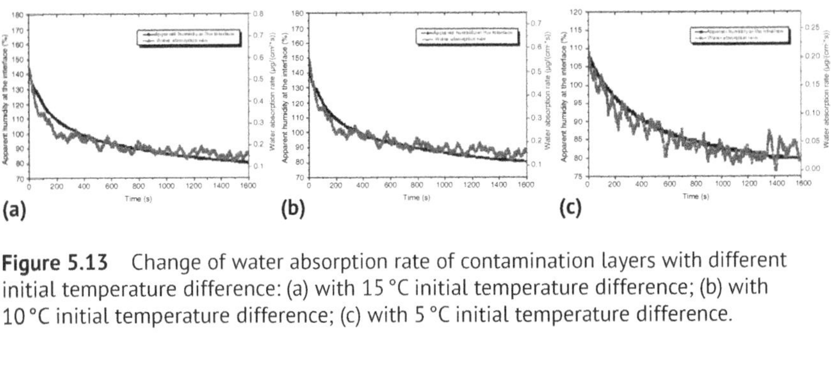

Figure 5.13 Change of water absorption rate of contamination layers with different initial temperature difference: (a) with 15 °C initial temperature difference; (b) with 10 °C initial temperature difference; (c) with 5 °C initial temperature difference.

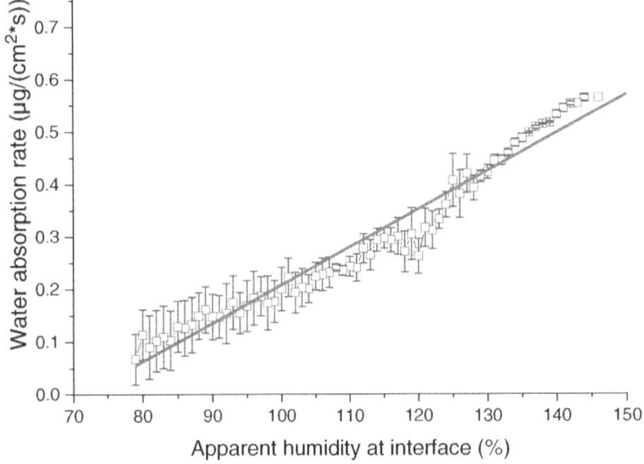

Figure 5.14 Relationship between apparent humidity at interface and water absorption.

When the apparent humidity exceeds 72.6%, the water absorption rate is positive, which means the contamination layers could absorb water molecules from the interface. Conversely, if the apparent humidity is below 72.6%, the water absorption rate of the contamination layers is negative, meaning that the contamination layer tends to release moisture back into the air. This phenomenon is attributed to the hygroscopic characteristics of NaCl. When NaCl is equilibrium with the surrounding atmosphere at 25 °C, the relative humidity of the air above the solution stabilizes around 75%.

5.4 Summary

This chapter investigated the influence of temperature difference between the insulator surface and ambient air on the leakage current and water absorption behavior of the contamination layer. Based on the experimental findings, the pollution warning model under low humidity was improved. The main conclusions are obtained as follows:

1. The maximum leakage current increases significantly with the humidity and is proportional to the initial temperature difference between the specimens and the atmosphere.
2. Regardless of the atmosphere dew point temperature, the water absorption rate of the insulator contamination layer increases with the temperature difference.
3. The water absorption rate of the contamination layer under temperature difference is proportional to the apparent humidity at the interface, which is higher than the ambient humidity.

By introducing the concept of apparent humidity at the interface, a more accurate pollution warning model for insulators under low humidity and temperature gradient conditions can be established. This improved model provides practical guidance for pollution monitoring in arid and semiarid regions, where conventional methods may fail to detect early-stage contamination risks.

Bibliography

Cao, B. and Wang, L. Wetting characteristics of artificial contamination on the hydrophobic surface, Eng. Fail. Anal., 2019, 100, 428–438, https://doi.org/10.1016/j.engfailanal.2019.02.049.

Cao, B., Mei, H., Wang, L. et al. The equivalent dissolved salt deposit density of contamination on surface of silicone rubber, Presented at the 2015 IEEE

Conference on Electrical Insulation and Dielectric Phenomena (CEIDP), 2015, 676–679.

Cao, B., Wang, L., Li, X. et al. Influence of partial electric arc on the contamination degree monitoring system in view of leakage current, IEEE Trans. Dielectr. Electr. Insul., 2018, 25 (4), 1545–1552. https://doi.org/10.1109/tdei.2018.006944.

Cao, B., Wang, L.M., and Yin, F.H. Measurement of saturated water absorption of the contamination layer deposited on insulator surface, IEEE Sens. J., 2019, 19 (22), 10804–10811. https://doi.org/10.1109/jsen.2019.2932192.

Chrzan, K.L., Vosloo, W.L., and Holtzhausen, J.P. Leakage current on porcelain and silicone insulators under sea or light industrial pollution, IEEE Trans. Power Del., 2011, 26 (3), 2051–2052. https://doi.org/10.1109/tpwrd.2011.2123470.

Crank, J. The Mathematics of Diffusion, Oxford University Press, 1979.

Dhahbi-Megriche, N. and Beroual, A. Time–frequency analyses of leakage current waveforms of high voltage insulators in uniform and non-uniform polluted conditions, IET Sci. Meas. Technol., 2015, 9 (8), 945–954. https://doi.org/10.1049/iet-smt.2015.0116.

Du, B.X., Liu, Y., Liu, H. J. et al. Recurrent plot analysis of leakage current for monitoring outdoor insulator performance (in English), IEEE Trans. Dielectr. Electr. Insul., 2009, 16 (1), 139–146. https://doi.org/10.1109/Tdei.2009.4784561.

Ghosh, R., Chatterjee, B., and Chakravor, S. A low-complexity method based on compressed sensing for long term field measurement of insulator leakage current, IEEE Trans. Dielectr. Electr. Insul., 2016, 23 (1), 596–604. https://doi.org/10.1109/tdei.2015.005458.

He, M., Li, H., Wang, J. et al. Superhydrophobic surface at low surface temperature, Appl. Phys. Lett., 2011, 98 (9), 093118. https://doi.org/10.1063/1.3558911.

Hsu, C.-C., Su, T.-W., Wu, C.-H. et al. Influence of surface temperature and wettability on droplet evaporation, Appl. Phys. Lett., 2015, 106 (14), 141602. https://doi.org/10.1063/1.4917291.

IEC TS 60815-1, Selection and Dimensioning of High-voltage Insulators Intended for use in Polluted Conditions - Part 1: Definitions, Information and General Principles, 2008.

Khafaf, N.A. and El-Hag, A.H. Prediction of leakage current peak value, Presented at the 2018 11th International Symposium on Mechatronics and Its Applications (ISMA), 2018, pp. 1–4.

Leclerc, M., Bouchard, R.P., Gervais, Y. et al. Wetting processes on a contaminated insulator surface (in English), IEEE Trans. Power App. Syst., 1982, 101 (5), 1005–1011. https://doi.org/10.1109/Tpas.1982.317352.

Menzel, D.H. Fundamental Formulas of Physics, Courier Corporation, 1960.

Palangar, M.F., Amin, U., Bakhshayesh, H. et al. Identification of composite insulator criticality based on a new leakage current diagnostic index, IEEE Trans. Instrum. Meas., 2021, 70, 1–10. https://doi.org/10.1109/TIM.2021.3096869.

Ruan, L., Han, G., Zhu, Z. et al. Spatial and temporal characteristics of insulator contaminations revealed by daily observations of equivalent salt deposit density, Sensors (Basel), 2015, 15 (2), 3023–3040. https://doi.org/10.3390/s150203023.

Wang, L., Cao, B., Guan, Z. et al. Influence of temperature difference on wetting characteristics of insulators, IEEE Trans. Dielectr. Electr. Insul., 2017, 24 (6), 3622–3629. https://doi.org/10.1109/tdei.2017.006813.

Wang, L., Cao, B., Mei, H. et al. Effects of natural contamination components on the surface conductivity under saturated moisture, IEEE Trans. Dielectr. Electr. Insul., 2017, 24 (5), 2945–2951. https://doi.org/10.1109/tdei.2017.006558.

Yang, L., Zhang, F., Wang, Z. et al. Evaluation of wetting condition and its effects on pollution flashover voltage of aerodynamic insulators (in English), IEEE Trans. Dielectr. Electr. Insul., 2016, 23 (5), 2875–2882. https://doi.org/10.1109/Tdei.2016.005821.

Zhang, Z., Jiang, X., Huang, H. et al. Study on the wetting process and its influencing factors of pollution deposited on different insulators based on leakage current, IEEE Trans. Power Del., 2013, 28 (2), 678–685. https://doi.org/10.1109/tpwrd.2012.2226758.

Zhao, C., Mei, H., Zhu, M. et al. Development of contamination flashover pre-warning system and analysis of operating experience, IEEE Trans. Dielectr. Electr. Insul., 2015, 22 (4), 2234–2241. https://doi.org/10.1109/tdei.2015.004472.

6

Characterization of Salt Fog Environments and Its Effect on the Outdoor Insulation

6.1 Introduction

The preceding chapters have systematically examined the wetting characteristics of pollution layers on insulator surfaces under various temperature and humidity conditions. However, one important scenario has not yet been addressed, namely the rapid salt fog commonly encountered in coastal regions. Such salt fog can not only induce rapid surface wetting of insulators within a short period of time, but also lead to a sharp increase in surface contamination severity, thereby significantly elevating the risk of pollution flashover.

According to IEC 60507 and IEEE Std. 4-2013, this type of contamination is classified as highly conductive liquid electrolytes. Under adverse weather conditions, particularly during salt fog and haze events, large numbers of fine water droplets or aerosol particles containing soluble salts are suspended in the atmosphere. The deposition of these soluble salts on insulator surfaces increases the amount of conductive contaminants, making flashover more likely under high humidity or wet conditions.

In current engineering practice, the pollution degree of external insulation is commonly evaluated using equivalent salt deposit density (ESDD) and non-soluble deposit density (NSDD). While ESDD shows a strong correlation with insulator flashover performance, it represents the cumulative effect of long-term atmospheric pollution and cannot accurately reflect the immediate influence of current pollution sources. Moreover, commonly used atmospheric pollution indicators, such as $PM_{2.5}$ and PM_{10}, characterize total particulate mass and do not directly correspond to the soluble salt content governing flashover behavior.

To more accurately characterize the impact of atmospheric pollution on external insulation performance, this chapter proposes the concept of **equivalent soluble salt concentration (ESSC)** in atmospheric air. The definition, measurement principle, and experimental validation of ESSC are presented, providing a new approach for assessing the flashover risk of insulators under salt fog and haze conditions.

6.2 ESSC

6.2.1 Definition and Measuring Principle

The concept of ESSC is similar to that of ESDD, but instead of quantifying salt contamination on solid surfaces, it characterizes the concentration of soluble salt contaminants in a unit volume of atmospheric air. ESSC, expressed in μg/L, is defined as the mass of NaCl that, when dissolved in demineralized or distilled water, produces the same volume conductivity as the soluble substances extracted from the pumped atmospheric air. This mass is then divided by the volume of the pumped atmospheric air, standardized to a temperature of 0 °C and a pressure of 101.3 kPa.

To determine ESSC, a known volume of atmospheric air is pumped through a container filled with measuring solution, allowing the airborne soluble salts to dissolve into the liquid. The temperature, volume, and volume conductivity of the solution are recorded both before and after air pumping. Given the pumped air volume, atmospheric pressure, and temperature known, the ESSC of the atmospheric air can be obtained and the calculation procedure is described in Figure 6.1.

The measured volume conductivity of the solution is converted to equivalent conductivity at 20 °C first by the formula from IEC 60815-1

$$\sigma_{20} = \sigma_\theta\left[1 - b(\theta - 20)\right] \tag{6.1}$$

where θ is the solution temperature (°C), σ_θ is the volume conductivity (S/m) at temperature θ, σ_{20} is the equivalent conductivity (S/m) at 20 °C, b is a factor which is a function of the temperature θ as denoted in Eq. (C2) of IEC 60815-1.

With given σ_{20}, the salinity of NaCl can be determined using the conductivity–salinity relationship also provided in IEC 60815-1:

$$S_a = (5.7\sigma_{20})^{1.03} \tag{6.2}$$

where S_a is the salinity of solution, expressed in kg/m^3.

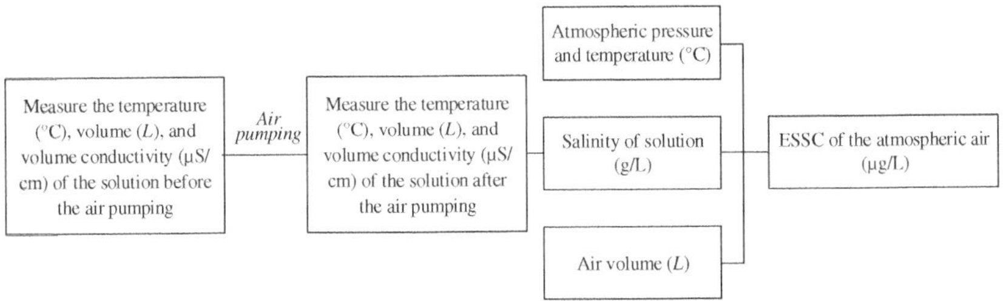

Figure 6.1 Calculation procedure of ESSC. *Source*: Cao et al. (2020)/with permission of John Wiley & Sons.

Assuming that the volume flow rate capacity of the air vacuum pump is Φ, then the volume of pumped air is:

$$V_{air} = \Phi \cdot \Delta t \tag{6.3}$$

where Δt is the pumping time of a pumping period.

The atmospheric pressure P_{air} and the temperature T_{air} of the atmosphere are measured using barometer and temperature sensors. According to ideal gas law, the volume of pumped air V_s at the standard temperature T_s and standard pressure P_s can be obtained by:

$$V_s = V_{air} \frac{P_{air}}{P_s} \frac{T_s + 273.15}{T_{air} + 273.15} \tag{6.4}$$

Finally, the ESSC of the atmospheric air can be calculated by:

$$\text{ESSC} = \frac{S_{a2}V_{water2} - S_{a1}V_{water1}}{V_s} \tag{6.5}$$

where S_{a1} and S_{a2} are the salinities of the solution before and after the air pumping, V_{water1} and V_{water2} are the volumes of the solution before and after the air pumping, V_s is the volume of pumped air (converted to the standard temperature and standard pressure).

6.2.2 Measuring Devices

Figures 6.2 and 6.3 display the workflow of the ESSC measurement platform and associated devices. The platform consists of an air extraction module for pumping a certain volume of air and a measurement module for measuring parameters related to the atmosphere and the aqueous solution. The air extraction module was composed of air inlet pipe, flow meter, air stone which was also called aquarium

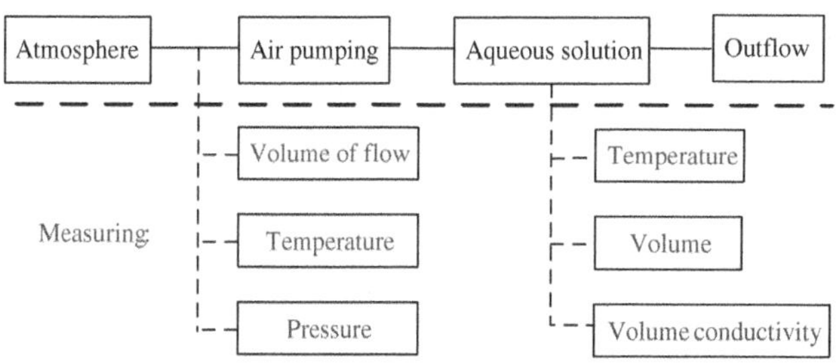

Figure 6.2 Structure of the ESSC measurement platform. *Source*: Cao et al. (2020)/with permission of John Wiley & Sons.

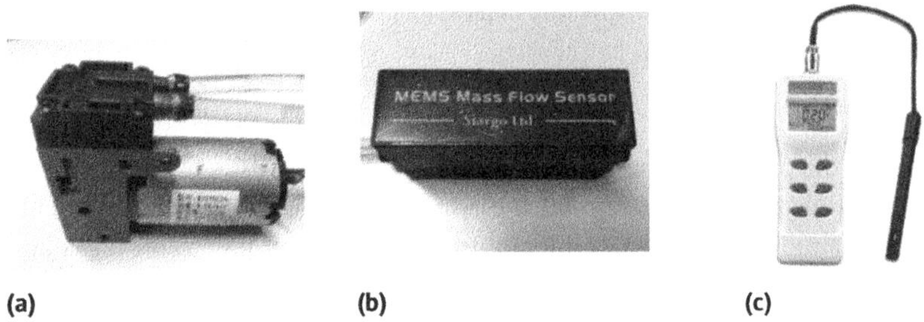

Figure 6.3 Measuring devices: (a) air pump: (b) mass flow sensor (FS4003): (c) conductivity meter (AZ8301). *Source*: Cao et al. (2020)/with permission of John Wiley & Sons.

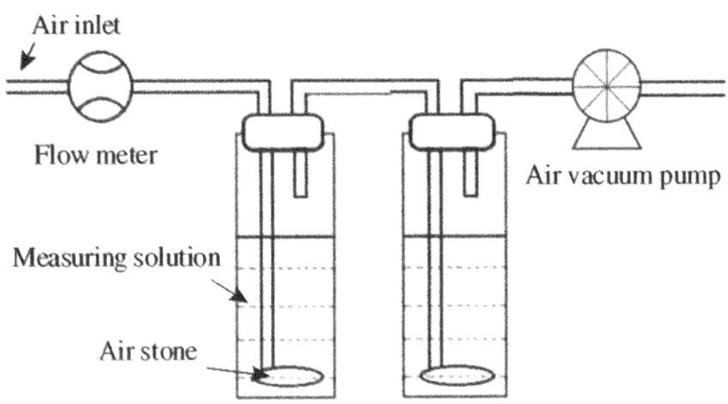

Figure 6.4 Schematic diagram of air extraction module. *Source*: Cao et al. (2020)/with permission of John Wiley & Sons.

bubbler, glass vial, and air pump. The air pump was from Shanghai Pengpu Fluid Technology Co. Ltd., and it had a maximum air extraction rate of 4.5 L/min at standard atmospheric pressure and room temperature.

The microelectromechanical (MEMS) mass flow sensor used was FS4003 from Siargo Ltd., the United States, featuring a flow range of 0~5 standard liters per minute. A handheld water quality meter AZ8301 was chosen to measure the volume conductivity and temperature of the aqueous solution. This instrument offers a conductivity accuracy of ±1.0% and a temperature precision of ±0.1 °C, with a measurement range for conductivity spanning from 0.1 μS/cm to 200 mS/cm.

Figure 6.4 presents a schematic of the air extraction module. The vials are sealed, and two pipes are inserted into each: one extending into the air above the solution, and the other immersed in the solution itself. The vial in the left is used to dissolve the soluble salt contaminants in the tested air. If the volume conductivity of the right vial is unchanged after the air extraction, it means that all the soluble salt contaminants in the tested air have been dissolved in the left vial. To extract the tested air into the solution, the air pump in the right side is turned on to generate an air pressure difference between inside the vial and the tested air environment. A flow meter in the left side is mounted to measure the volumetric flow rate of air extracted into the solution.

To facilitate the soluble salt contaminants of the pollutants in the tested air to be completely dissolved in the solution, the following measures were taken. First, the pipe end in the solution was connected to an air stone. The air stone at the vial bottom can gradually diffuse air into the tank and therefore eliminate the large bubbles of conventional air filtration systems and increase the contact area between air bubbles and the solution. Secondly, a slender vial was chosen. It can increase the time that the air bubble would taken to reach the surface of the solution.

To measure the ESSC of atmosphere accurately, it is critical to select an appropriate volume of water for the measurement, as well as an adequate introduction rate of measuring air. It is obvious that the soluble salt contaminants in the atmosphere are more likely to dissolve completely in water at large volume. On the other hand, however, too much water will result in small volume conductivity change of the solution and decrease the measurement sensitivity when the pollution in the atmosphere is light.

According to the literature on the conductivity of fog in the natural environment, the volume conductivity of fog-haze was generally 500–1,000 μS/cm in ordinary weather and could reach 2,000–3,000 μS/cm in severe condition. Moreover, a dynamic fog collector was used to measure the liquid water content (LWC) during fog-haze days. The measured LWC in the fog-haze was between 0.01 and 3.0 g/m^3, and the maximum LWC was 4.83 g/m^3 under dense foggy weather condition. In the tests followed in this chapter, the air extraction rate was set at 2.7 L/min while the volume of measuring solution was 100 mL. In all the tests in this chapter, the volume conductivity of the right vial was unchanged after the air extraction. Therefore, it can be concluded that the air extraction rate and the measuring solution volume selected were reasonable.

6.2.3 Experimental Arrangement for the Salt Fog Test

Salt fog test is normally carried out in the climate chamber. In this section, the fog was artificially simulated with water conductivity ranging from 1,000 μS/cm to 5,000 μS/cm by a salt sprayer. In this book, the ESSC of the salt fog environment was measured in the fog-haze chamber by the above-described ESSC measurement platform.

6.2.3.1 Experiment Setup

The artificial salt fog test was conducted in a fog-haze chamber, which was fabricated specially for this purpose. The chamber, as shown in Figure 6.5, is made of plexiglass with an outer diameter of 100 cm, an inner diameter of 97 cm, and a

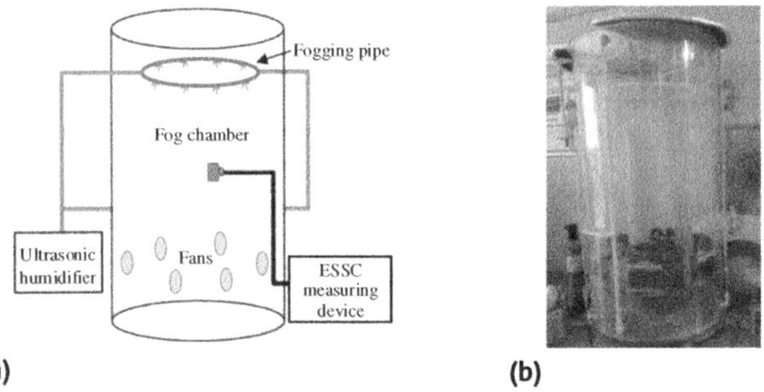

Figure 6.5 Fog-haze chamber used for the artificial salt fog test: (a) the schematic diagram of the fog-haze chamber; (b) appearance of the fog-haze chamber. *Source*: Cao et al. (2020)/with permission of John Wiley & Sons.

height of 200 cm. The effective test space is between a height of 65 cm and 165 cm from the chamber floor.

Fog was channeled into the fog-haze chamber via the hose connected to the chamber and spread within the chamber via the perforated circular plastic tube. Located at a height of 190 cm, the circular plastic tube was coaxial with the fog-haze chamber. Six fans were placed uniformly around the inner wall of the fog-haze chamber at a height of about 30 cm from the chamber floor to drive the air to the central area.

During each test, salt fog was generated by an ultrasonic humidifier RN-12 from Taizhou Ryan, China. The ultrasonic humidifier had a max mist/fog output of 800 mL/h and it consisted of a transducer (a high-frequency oscillator made of piezoelectric material) and an excitation circuit. The transducer was installed at the bottom of flume and was in direct contact with water. Under the action of the excitation circuit, high-frequency vibrations of 1.7 MHz were generated by the transducer to produce ultrasonic waves. When the water was acted by the ultrasonic waves, uniform foggy micron-level water droplets were generated and escaped from the water surface. The exit of the humidifier was connected to the hose and the circular plastic tube. Therefore, the fog can fill the chamber uniformly.

6.2.3.2 Test Procedures

In the artificial salt fog test, salt fog was prepared with NaCl solution to simulate the soluble salt contaminants in the atmospheric air. The test procedures are as follows:

1. Prepare the NaCl solution with desired volume conductivity and pour it into the ultrasonic humidifier. Install the air extraction module of the ESSC measurement platform.

2. Connect the exit of ultrasonic humidifier to the fog-haze chamber, and then turn on the humidifier to fill the chamber with fog. After about 30 minutes, the fog concentration in the chamber reaches a steady state.
3. Keep the fog generation rate constant and activate the air pump to extract air for a certain period of time. The temperature, volume, and volume conductivity of the solution are measured before and after each air pumping.
4. The average ESSC of the artificial fog in the fog-haze chamber is calculated by the method mentioned above.
5. After the test, turn off the humidifier and keep the fans in the fog-haze chamber running to expel the artificial salt fog out of the chamber.

To minimize the impact of air extraction on internal pressure and salt concentration in the chamber, the air pump was operated for five minutes in each 30-minute test cycle. The extracted extraction volume of the salt fog in each measurement cycle was approximately 0.014 m^3, which was less than 2.5% of the total effective volume of the fog-haze chamber. Therefore, the influence of air extraction on the air pressure in the chamber is negligible.

6.2.3.3 Selection of Water Conductivity and Fog Concentration

In the artificial salt fog test, the salt fog was simulated by NaCl solution with volume conductivity ranging from 1,000 and 5,000 μS/cm. The fog generation rate by the ultrasonic humidifier was chosen according to the volume of fog-haze chamber. In the tests, the fog generation rates were set to 140 and 310 mL/h based on the parameters of ultrasonic humidifier and the volume of fog-haze chamber.

6.3 Results

6.3.1 ESSCs at Various Fog Water Conductivities

The fog generation rate of the humidifier was maintained constant throughout each test. For the ESSC measurement, water with four different volume conductivities, as listed in Table 6.1, was used to generate the artificial fog, respectively.

As shown in Table 6.1, the ESSC of the fog is between 22.48 and 31.92 μg/L when the volume conductivity is 1183 μS/cm; between 35.34 and 59.81 μg/L when the volume conductivity is 2109 μS/cm; between 50.88 and 75.83 μg/L when the volume conductivity is 3,380 μS/cm; and between 84.65 and 98.81 μg/L when the volume conductivity is 4437 μS/cm. The average ESSCs are also listed at the bottom of the Table 6.1. It can be seen that the ESSC increases with an increase of volume conductivity. With the same fogging rate and environmental conditions, the fog concentration inside the fog-haze chamber is almost identical. Therefore, the ESSC of the fog inside the chamber depends on the water conductivity.

Table 6.1 Measured ESSC of the fog generated by water with different volume conductivities (fogging rate: 310 mL/h).

	Measuring time interval	Measured ESSC of the fog generated with different volume conductivities (μg/L)			
No.	**Start-stop (min)**	**1,183 μS/cm**	**2,109 μS/cm**	**3,380 μS/cm**	**4,437 μS/cm**
1	30–35	31.92	55.88	65.34	93.17
2	60–65	25.15	59.81	71.86	98.81
3	90–95	27.47	51.49	75.83	95.85
4	120–125	24.54	48.02	56.18	93.08
5	150–155	26.26	46.53	50.88	84.65
6	180–185	22.48	41.36	56.48	92.32
7	210–215	23.62	37.91	67.59	85.45
8	240–245	25.91	35.34	64.29	88.45
9	270–275	24.82	38.65	68.4	87.84
Avg.		25.80	46.11	64.09	91.07

The measured ESSC fluctuates somewhat over time with the same volume conductivity. Except for the possible variations of airborne soluble salts in the fog-haze chamber, it may also attribute to the air pump used in the tests. It is very difficult to keep the air extraction rate constant even with the help of a pressure regulator valve which can regulate and stabilize the air pressure.

6.3.2 ESSCs at Various Fog Water Concentrations

During the artificial salt fog test, the fog concentrations can be varied by changing the fogging rate of the humidifier. In the present tests, two fogging rates of the humidifier, namely 310 and 140 mL/h, were tested.

With different fogging rates, the injection rate of the fog into the chamber and the fog concentration inside the chamber differ. Visibility is a common parameter to characterize the fog concentration in meteorology. Figure 6.6 shows the appearance of the fog-haze chamber after a steady state has been reached at two different fogging rates (140 mL/h on the left and 310 mL/h on the right). It is obvious that the visibility of the right picture is much lower than that of the left picture. When the fogging rate is 310 mL/h, the fog outlet tube at the top and the fan at the bottom are more blurred.

While keeping the humidifier's fogging rate constant at 140 mL/h, the measured ESSCs of the salt fog inside the fog-haze chamber at three volume conductivities are listed in Table 6.2. As shown in the table, when the fog generation rate

Figure 6.6 Appearance of the fog-haze chamber at two different fogging rates. *Source*: Cao et al. (2020)/with permission of John Wiley & Sons.

Table 6.2 Measured ESSC of the fog generated by water with different volume conductivities (fogging rate: 140 mL/h).

	Measuring time interval	**Measured ESSC of the fog generated with different volume conductivities (μg/L)**		
No.	**Start-stop (min)**	**1,378 μS/cm**	**4,041 μS/cm**	**5,001 μS/cm**
1	30–35	19.24	26.88	47.77
2	60–65	11.12	33.64	52.25
3	90–95	12.51	27.55	62.29
4	120–125	19.8	34.35	45.08
5	150–155	15.95	34.29	51.26
6	180–185	17.23	35.33	51.19
7	210–215	15.79	26.58	62.81
8	240–245	15.67	31.39	62.03
9	270–275	22.45	39.31	55.55
Avg.		16.64	32.15	54.47

of the ultrasonic humidifier is 140 mL/h, the ESSC inside the chamber is between 11.12 and 22.45 μg/L when the volume conductivity is 1378 μS/cm; between 26.58 and 39.31 μg/L when the volume conductivity is 4041 μS/cm; and between 45.08 and 62.81 μg/L when the volume conductivity is 5,001 μS/cm.

Figure 6.7 shows a comparison of the relationships between the average ESSC and the volume conductivity at two fogging rates. At both fogging rates, the average

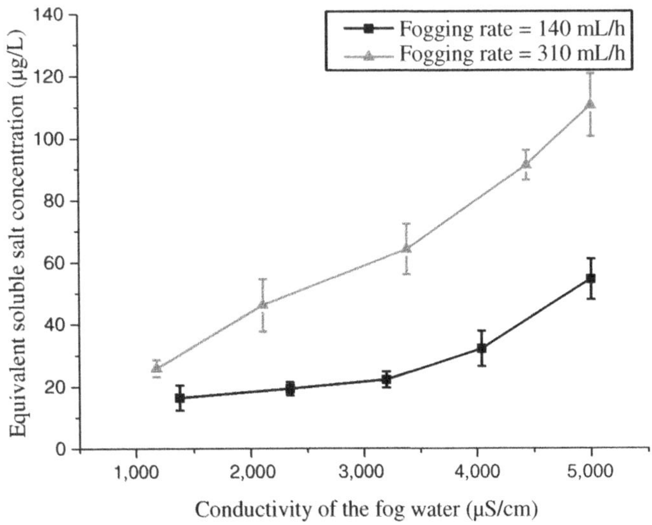

Figure 6.7 Comparison of ESSC and the volume conductivity of the fog water with two fogging rates.

ESSC increases with an increase of volume conductivity. Moreover, with the same volume conductivity of the water, the ESSC of the salt fog at the fogging rate of 310 mL/h is much higher than that at the fogging rate of 140 mL/h. Obviously, ESSC can reflect the combined effect of volume conductivity of water and the fog concentration in the salt fog. At the same fog concentration, a higher volume conductivity leads to a larger ESSC and at the same volume conductivity, a higher fog concentration also results in a larger value of ESSC.

6.3.3 On-site Measurement Tests

The developed ESSC measurement platform was also used for in-field measurements in Nanjing, Jiangsu Province, China. The typical pollution sources around the measurement point were industrial dust and waste gas, as shown in Figure 6.8.

In order to prolong the service life of the air pump and ensure the accuracy of the measurement results, the air pump alternated between running for 15 minutes and halt for 45 minutes when measuring the ESSC. Two weeks later, the volume conductivity of the solution was measured and the ESSC was calculated. It was found that the ESSC of the atmospheric air is 0.261 μg/L, which was about 0.016 times of that in the artificial salt fog chamber when the fogging rate was 140 mL/h and conductivity fog was 1378 μS/cm.

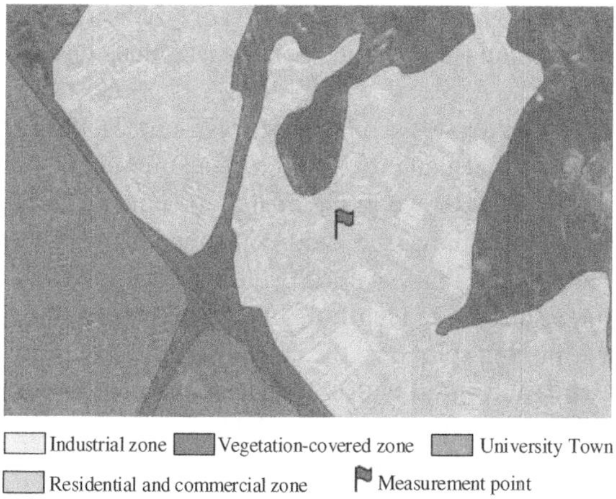

Figure 6.8 Location of the measurement point and the typical pollution sources nearby. *Source*: Cao et al. (2020)/with permission of John Wiley & Sons.

Figure 6.9 Contamination deposition test platform: (a) shutter box placed on the stand; (b) glass specimens sited in the box. *Source*: Cao et al. (2020)/with permission of John Wiley & Sons.

To study the relationship between ESSC and the insulator surface contamination, glass test specimens with dimension of 4 × 3 × 0.5 cm (length × width × thickness) were also arranged at the measurement point. To avoid the contamination being washed by natural rainfall, the specimens were placed in a shutter box, as shown in Figure 6.9. After two weeks of natural contamination, the contamination on the specimen surface was analyzed.

Due to the short exposure time and light contamination, it was difficult to obtain the ESDD of the contamination layer by the procedure given in IEC 60815-1. At the end of the test, the contamination on the specimen surface was scrubbed and dissolved into 20 mL deionized water and the soluble salt ion compositions and their concentrations in the solution were analyzed by an ion chromatography. The major ion compositions and their surface densities are listed in Table 6.3.

To compare with the contamination deposited on the specimen surface, the ion compositions of the solution used to calculate the ESSC of the atmospheric air was also measured after air pumping, as shown in Figure 6.10. It could be found that in the in-field measurements in Nanjing, the major ion components of the solution made by the ESSC method and that obtained by dissolving contamination scrubbed from the specimen surface are identical, and the mass percentage of each ion is almost the same. In both solutions, the cation ion K^+ account for almost half of the ion mass while the anion with the highest mass percentage is Cl^-. Therefore, it can be inferred that the measured soluble salt in the atmospheric air at the measurement point would also be the main source of the soluble salt deposited on the insulator surface nearby. The ESSC measured by the method proposed in this chapter could be used as a parameter for the prediction of insulator surface contamination. As it is a very complicated problem, their relationship needs to be further studied in the future.

Table 6.3 Ion compositions and their surface densities of the contamination deposited on the glass specimen surface.

Ion type	Na^+	K^+	Ca^{2+}	Mg^{2+}	Cl^-	SO_4^-
Ion density ($\mu g/cm^2$)	0.647	3.653	0.955	0.379	1.395	0.764

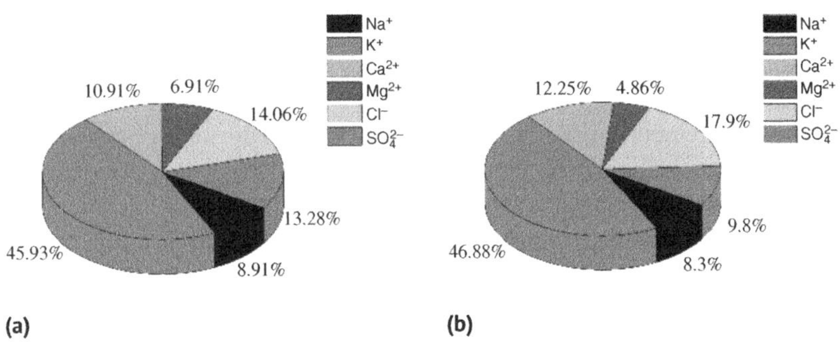

Figure 6.10 The major ion composition of the solutions and their mass percentages: (a) ESSC measuring solution; (b) the solution from the specimen surface. *Source*: Cao et al. (2020)/with permission of John Wiley & Sons.

6.4 Comprehensive Model for Predicting Flashover Voltage in the Salt Fog Environments

6.4.1 Test Platform Overview

The research was done in a small spray chamber with controllable temperature and humidity. The whole test system includes the spray chamber, a high-voltage power source, the monitoring equipment of leakage current, and the control software of the system.

6.4.1.1 Small Spray Chamber with Controllable Temperature and Humidity

The small spray chamber is composed of a plexiglass cylinder which has a diameter of 85 cm and is 120 cm in height. It is equipped with an ultrasonic mist generator, dehumidifying air exhauster, chassis fan cycle, heating wires that are uniformly distributed on the walls of the chamber, temperature and humidity sensor and supervision system.

In the spray chamber, the relative humidity can be automatically adjusted from ambient levels up to 99%. The temperature can be controlled automatically from the environment temperature to 30 °C higher, which should be less than 60 °C.

It has been proved by the experiment that the distribution gradient in the chamber is very small, as the temperature gradient $< \pm 1$ °C and the humidity gradient $< \pm 3\%$, which is within the desired range of the technical parameters of the chamber. It also means that in the spray chamber, the impact from the sample hanging position on the wetting characteristics can be ignored.

6.4.1.2 Conductivity Fog Producer

In this chapter, the conductivity fog was produced by ultrasonic mist filter, with conductivity solution. By condensing the fog and measuring its conductivity, it has been proved that this method can create the conductivity fog successfully.

Previous studies have pointed out that when at the same conductivity, the component of the fog has no effect on the flashover voltage. Therefore, the NaCl was chosen to create the solution, using deionized water and ignoring the impact of undissolved substance in the air.

6.4.1.3 Other Components of the System

The test platform with controllable temperature and humidity also includes other components as power system, leakage current measuring system, and monitoring software.

The power system consists of a voltage regulator, a voltage transformer, a voltage divider, a test console, and voltage measurement instruments. The capacity of the transformer is 100 kVA, while the maximum output voltage is 50 kV and the rated output current is 2 A. It can be used in the test on line insulators below 35 kV.

The leakage current measuring system is composed of leakage current sensor, data acquisition card, power protector, and personal computer. The leakage current sensor was developed by the lab, which is kind of contact resistive-type sensor. It has overvoltage and overcurrent protect device to get rid of introducing high voltage into the measuring system when the dielectric breakdown occurs. The measurement range of the system is from 0.5 mA to 5 A, while the total error is controlled within plus or minus 3%.

The monitoring software is developed based on LabView language, including the measuring window of test voltage and leakage current, the statistics of leakage current's characteristics, the temperature in the small spray chamber, the measurement and control interface of humidity.

6.4.1.4 Sample of the Test

The insulator model used in this study was the XP-70 type.

6.4.1.5 Test Method

The quantitative brushing method was used in the test, with kaolin as undissolved substance and NaCl as dissolved substance.

Put one insulator hanging in the spray chamber, when the contamination is fully moisture, use the evenly voltage-rising method to find the flashover voltage. There are three samples in each condition, and each sample gets flashover for three times. Each flashover should be done 10 minutes after the last one, in order to ensure that the dry area formed possibly has been damped fully. The average voltage represents the flashover voltage, which requires that the difference between each datum and the mean value less than 15%.

6.4.2 Results and Analysis

6.4.2.1 Comparison Between Conductive Fog and Clean Fog

To investigate the impact of conductivity fog on the flashover voltage, the fog conductivity was set to 2,000 μS/cm and the results were compared with those under clean fog conditions. The salt deposit density divers from 0.05 to 0.25 mg/cm^2, while the dust density is at the same level at 1.0 mg/cm^2.

As shown in Figure 6.11, when the ESDD and NSDD are at the same level, the conductivity fog decreases the flashover voltage. As the ESDD grows, the flashover voltage has the same kind of saturated trend. But there are some differences under different ESDD: at low ESDD, the reduction is significant—up to 28.2%; at higher ESDD, the decrease is limited to around 13.6%.

Studies have shown that the root cause of the conductivity fog's function is the accumulation of salt density on the insulator surface, and the discharge is still

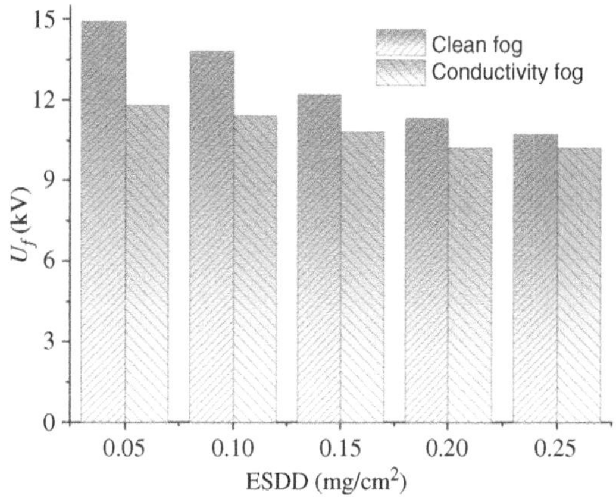

Figure 6.11 Graph of flashover voltage vs. different ESDD under clean fog and conductivity fog.

on surface instead of air gap breakdown. In this research, the flow and moisture time were kept the same, with the same NSDD, which resulted in the same level of water gathered on the insulator surface. The declining tendency of the flashover voltage decrease is then mainly due to the limited solubility of the salt. When the insulator is clean, the salt on surface has fully dissolved after the moisture. The conductivity fog can cause a significant increase of the dissolved salt, which occurs to be the increase of surface conductivity. But when the ESDD is high, which means the salt stone has already near its limit, the fog then can barely increase the surface conductivity. This explains the diminishing effect of conductivity on voltage reduction as ESDD increases.

In summary, at the areas with light pollution, such as coastal areas, it's important to specially consider the risk coming from the conductivity fog. Because of the neglect of them, with the low insulation solution, it may be easier to get flashover when the bad weather happens suddenly. As to the areas with heavy pollution, with the lower impact, the insulation solution may be enough to get rid of the accident caused by the conductivity fog.

6.4.2.2 Relationship Between Flashover Voltage and Fog Conductivity

Under clean fog conditions, the relationship between flashover voltage U_f and ESDD can be described as:

$$U_f = A \cdot S^{-a} \tag{6.6}$$

Based on experimental data, this was fitted as:

$$U_f = 8.1052 \cdot S^{-0.2119} (\text{kV}) \tag{6.7}$$

as illustrated in Figure 6.12.

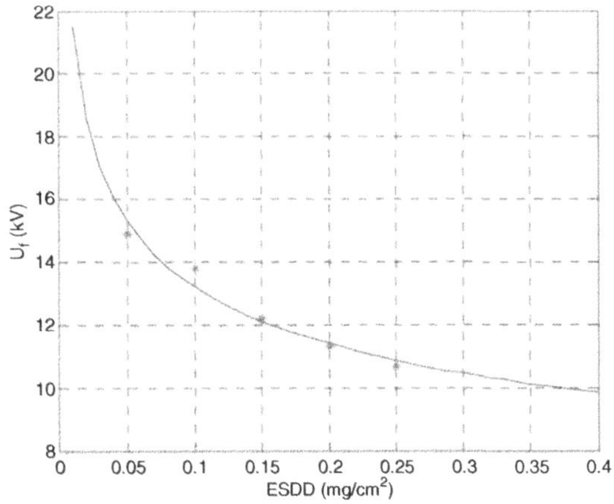

Figure 6.12 Flashover voltage vs. ESDD under conductivity fog (2,000 μS). *Source*: Wang et al. (2015)/with permission of IEEE.

The influence of conductive fog was treated as equivalent to an additional salt deposit, modifying the formula to:

$$U_f = A \cdot (S + S_f)^{-a} \tag{6.8}$$

where S in the equation stands for the original ESDD before the conductivity fog comes, and the S_f stands for the result of conductivity fog. The two elements are combined by the same flashover voltage. But with the research data from this test, it turns out to be not effective, as the ESDD turns out irregular.

Further attempts were made to correlate fog conductivity with the fitting coefficient, treating ESDD as an independent variable, but this approach also failed to yield a consistent model.

Because the conductivity fog affects the flashover by increasing the salt density, referring to the expression of salt density and flashover voltage, the experiment data was fitted multiply. The result from the joint effects can be concluded as:

$$U_f = 9.58 \cdot S^{-0.161} \cdot {\gamma_{20}}^{-0.0214} \tag{6.9}$$

where γ_{20} stands for the fog conductivity reduced to 20 °C.

As illustrated in Figure 6.13, both increasing ESDD and fog conductivity contribute to a decline in flashover voltage. The surface plot confirms the saturation trend predicted by the fitted model, demonstrating the interactive effect of salt density and atmospheric conditions.

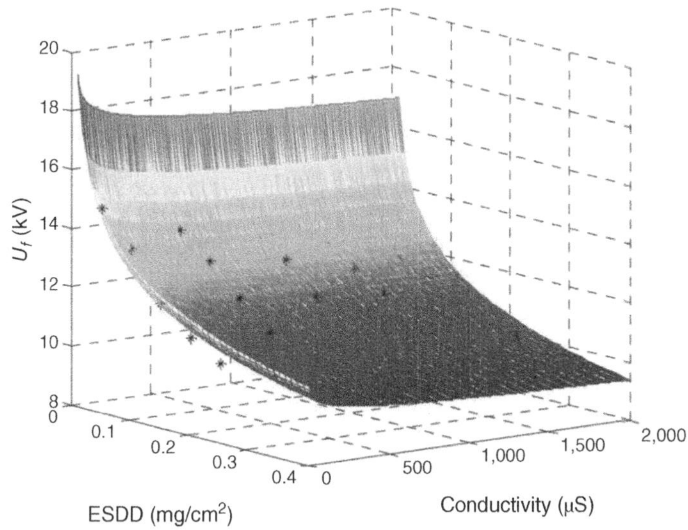

Figure 6.13 Flashover voltage vs. ESDD and fog conductivity. *Source:* Wang et al. (2015)/ with permission of IEEE.

The 3D surface plot shows that flashover voltage decreases nonlinearly with both increasing salt contamination (ESDD) and fog conductivity, validating the joint effect captured in Eq. (6.9).

As shown in Figure 6.14, with the increase of ESDD or fog conductivity, the flashover voltage will both decrease as well as in a saturated trend, which fits well with the actual situation.

This model yielded a residual of 0.0019 and a goodness-of-fit (R^2) of 0.8563. After removing two outlier data points identified via residual confidence intervals, the improved model is:

$$U_f = 8.94 \cdot S^{-0.188} \cdot \gamma_{20}{}^{-0.0179} \tag{6.10}$$

The updated fit achieves a lower residual of 0.0011 and improved R^2 of 0.9198. The increased exponent on ESDD further confirms that ESDD has a stronger effect on flashover voltage than fog conductivity.

The fitting expression can be used in the forecasting of flashover voltage in the warning system. With the data of fog conductivity, from meteorological department or measuring equipment, the flashover voltage can be revised. When the fog is coming, the risk of flashover can be known. It can also get the conductivity which will cause flashover under operating voltage. If this kind of weather appears commonly, the insulator solution should be improved immediately to avoid the accident.

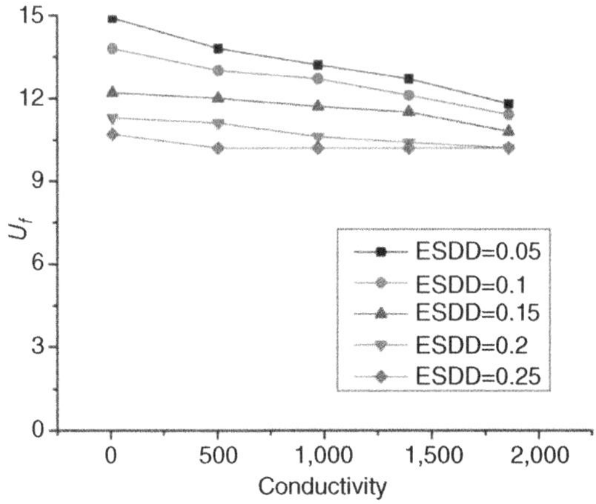

Figure 6.14 Flashover voltage vs. fog conductivity with different ESDD on insulators. *Source*: Wang et al. (2015)/with permission of IEEE.

6.5 Summary

ESDD, as a static parameter, cannot predict the pollution degree alone while other environmental pollution parameters such as PM2.5, PM10, and API do not distinguish the soluble salts from insoluble components. For the safe operation of high-voltage power systems, a new parameter ESSC was proposed in this chapter.

1. The proposed ESSC serves as a real-time indicator of soluble salt contamination in the ambient air, which is closely correlated with the surface pollution level of insulators. The concept, definition, and measurement principle of ESSC were presented in detail. By quantifying ESSC, it is possible to simultaneously evaluate both the proportion of soluble salt within total pollutants and the overall contamination concentration.
2. Based on the measurement principle, a dedicated ESSC testing platform was developed. This system was used to monitor salt fog conditions inside a test chamber. It was experimentally verified that ESSC values increase proportionally with both the fog concentration and the conductivity of the water used to generate the fog.
3. Field tests were also conducted using the ESSC platform and glass specimen collectors. In an industrial area of Nanjing, the atmospheric ESSC was measured to be 0.261 μg/L. Ion chromatography revealed that the main ionic components and their mass ratios in the specimen-washed solution were nearly identical to those in the sampled atmospheric air. This confirms that, over short periods, the soluble pollutants depositing on surfaces reflect the actual composition of the air.

ESDD represents a cumulative effect over years, while ESSC captures the real-time condition of soluble salt concentration in the air. Given that pollution on insulator surfaces accumulates progressively, it is conceivable to predict future ESDD by combining periodic ESDD measurements with continuous ESSC monitoring. However, due to the complexity of pollution dynamics, more comprehensive research is required to establish an accurate predictive model.

High-conductivity fog significantly reduces the flashover voltage compared to clean fog conditions. The reduction becomes more pronounced at lower ESDD levels, with a diminishing effect as ESDD increases—exhibiting a saturation trend. This effect is especially critical for relatively clean insulators.

Given the similar physical mechanism between conductivity fog and surface salt deposits, a unified flashover voltage expression was developed based on both fog conductivity and ESDD. The fitted model demonstrates high accuracy and can be integrated into early warning systems for insulation failure. With this model, flashover voltage can be dynamically corrected according to real-time weather data. If fog-induced flashovers are found to occur frequently, insulation configurations should be promptly upgraded to ensure safe system operation.

Bibliography

Cao, B., Wang, L., and Yin, F. Measurement of saturated water absorption of the contamination layer deposited on insulator surface, IEEE Sens. J., 2019, 19 (22), 10804–10811.

Cao, B., Yin, F., Yang, D. et al. Characterisation parameter of atmospheric pollutant concentration for external insulation, IET Gener. Transm. Distrib., 2020, 14 (20), 4547–4552.

Chrzan, K., Stec, C., and Farzaneh, M. Probes for spot measurement of surface conductivity on polluted insulators, IET Sci. Meas. Technol., 2007, 1 (5), 295–300.

Douar, M., Beroual, A., and Souche, X. Degradation of various polymeric materials in clean and salt fog conditions: measurements of AC flashover voltage and assessment of surface damages, IEEE Trans. Dielectr. Electr. Insul., 2015, 22 (1), 391–399.

Farzaneh, M. and Chisholm, W.A. Insulators for icing and polluted environments, IEEE Press Series on Power Engineering, Wiley, 2009.

IEC 60507-2013, Artificial Pollution Tests on High Voltage Insulators to Be Used on AC Systems, 2013.

IEC 60815-1, Selection and Dimensioning of High-voltage Insulators Intended for Use in Polluted Conditions - Part 1: Definitions, Information and General Principles, 2008.

IEEE Std. 4-2013, IEEE Standard Techniques for High-Voltage Testing, 2013.

Ilhan, S. and Cherney E. Comparative tests on RTV silicone rubber coated porcelain suspension insulators in a salt-fog chamber, IEEE Trans. Dielectr. Electr. Insul., 2018, 25 (3), 947–953.

Jiang, X., Dong, B., Hu, Q. et al. Effect of ultrasonic fog on AC flashover voltage of polluted porcelain and glass insulators, IEEE Trans. Dielectr. Electr. Insul., 2013, 20 (2), 429–434.

Jiang, X., Wu, Q., Zhang, Z. et al. Estimation of rime icing weight on composite insulator and analysis of shed configuration. IET Gener. Transm. Distrib., 2018, 12, 650–660.

Leng, C., Roberts, J., Zeng, G. et al. Effects of temperature, pH, and ionic strength on the Henry's law constant of trimethylamine, Geophys. Res. Lett., 2015, 42, 3569–3575.

Liu, Y., Du, B., and Farzaneh, M. Self-normalizing multivariate analysis of polymer insulator leakage current under severe fog conditions, IEEE Trans. Power Del., 2017, 32 (3), 1279–1286.

Stefenon, S., Grebogi, R., Freire, R. et al. Optimized ensemble extreme learning machine for classification of electrical insulators conditions, IEEE Trans. Ind. Electron., 2020, 67 (6), 5170–5178.

Sudalai Shunmugam, S., Vasudev, N., Ravi, K.N. et al. Applicability of dimensional analysis for the prediction of pollution performance of insulators: an experimental study. IET Gener. Transm. Distrib., 2017, 11, 1319–1324.

Sun, Y., Tu, Y., Wang, C., et al. Contamination and AC pollution flashover characteristics of insulators under fog-haze environment, IEEE 11th International Conference on the Properties and Applications of Dielectric Materials (ICPADM), 2015, pp. 596–599.

Wang, K., Zhao, C., Zhu, M. et al. Analysis on flashover performance of polluted porcelain insulators in high conductivity fog environment, IEEE Conference on Electrical Insulation and Dielectric Phenomena (CEIDP), 2015, pp. 193–196.

Wang, L., Cao, B., Mei, H. et al. Effects of natural contamination components on the surface conductivity under saturated moisture, IEEE Trans. Dielectr. Electr. Insul., 2017, 24 (5), 2945–2951.

Wardman, J., Wilson, T., Hardie, S. et al. Influence of volcanic ash contamination on the flashover voltage of HVAC outdoor suspension insulators, IEEE Trans. Dielectr. Electr. Insul., 2014, 21 (3), 1189–1197.

Yin, F., Jiang, X., and Farzaneh, M. Electrical performance of composite insulators under icing conditions, IEEE Trans. Dielectr. Electr. Insul., 2014, 21 (6), 2584–2593.

Yuan, Q., Li, W., Zhou, S. et al. Integrated evaluation of aerosols during haze–fog episodes at on regional background in North China plain. Atmos. Res., 2015, 156, 102–110.

Zhang, Z., Jiang, X., Huang, H., et al. Study on the wetting process and its influencing factors of pollution deposited on different insulators based on leakage current, IEEE Trans. Power Deliv., 2013, 28 (2), 678–685.

Zhao, Q., Meng, X., and Chen G. Chemical characteristics of fog water in Chongqing during the tenth five-year plan, J. Southwest Univ. Nat. Sci. Ed., 2007, 29 (8), 146–149.

Zhou, J., Xing, Z., and Deng, J. Characterizing and sourcing ambient PM2.5 over key emission regions in China I: water-soluble ions and carbonaceous fractions, Atmos. Environ., 2016, 135, 20–30.

7

Effect of Hydrophobicity on the Contamination Wetting

7.1 Introduction

The second part of this book, covered in Chapters 7–10, focuses on the hydrophobic properties of insulation materials and their influence on contamination wetting and pollution flashover behavior. Silicone rubber, which is widely used in composite insulators, exhibits strong intrinsic hydrophobicity and hydrophobic transfer properties. Under wetting conditions, water tends to form discrete droplets rather than continuous water films on the insulator surface. This characteristic effectively increases the flashover voltage and suppresses the occurrence of pollution flashover. This section examines the influence of varying levels of surface hydrophobicity on flashover performance, and how these effects are modulated by wetting conditions.

Existing studies on wetting mechanisms and salt dissolution behavior on hydrophobic insulation surfaces remain limited. Previous investigations have shown that leakage current on hydrophobic surfaces during natural wetting is generally small, intermittent, and highly unstable. The cleaning effect of rainfall on soluble salts deposited on silicone rubber surfaces with different hydrophobic levels has also been studied, leading to the proposal of the concept of effective ESDD to characterize the amount of dissolved salt retained on composite insulator surfaces after irrigation. Nevertheless, the wetting characteristics of hydrophobic surfaces have not yet been sufficiently clarified.

It investigates how different levels of surface hydrophobicity and material characteristics influence flashover behavior and how these effects are modulated by other factors.

It has been proved that the wetting degree of contamination layer significantly influences the flashover voltage of insulator string. If the wetting duration is too short, the contamination layer remains insufficiently wetted, resulting in relatively high flashover voltage. This condition fails to accurately reflect differences in hydrophobic performance among specimens. Conversely, if the wetting time is too long, the small water drops on the contamination layer continue to aggregate into larger drops or water film, which may even lead to loss of contamination solution.

Due to the hydrophobicity and hydrophobic transfer property, the wetting process on composite insulators is very complicated. Currently, research on the wetting and contaminant dissolution processes for composite insulators remains limited. The characteristics of leakage current on the hydrophobic surface were studied, it was found that the leakage current on the hydrophobic surface during natural wetting was very small, intermittent and extremely unstable. Cleaning effect of rainfall to the soluble salts on the silicone rubber with different hydrophobicity was studied and the concept of effective ESDD of composite insulators was proposed to characterize the dissolved salt which is retained on the insulator surface after irrigation. In spite of present achievements, investigations on the wetting characteristics of hydrophobic surface were still insufficient.

Additionally, to scientifically select the type and string length of insulators with organic insulation material in various environments, many researchers have conducted artificial contamination tests under various conditions to study the effects of insulator suspension forms, shed shapes, altitudes, and other parameters on the pollution flashover voltage. In the artificial contamination test, visual inspection is generally used to judge the degree of wetness of the contaminated layer, however, due to the lack of sufficient researches on the wetting characteristics of contamination on the hydrophobic surface, it is difficult to evaluate when the pollution layer is saturated absorption scientifically.

To investigate the wetting behavior of artificial contamination on the surface of composite insulators, this chapter measured the salinity of water droplets and the surface conductivity of contamination layers during the wetting process. In addition, high-magnification microscopy was employed to observe the morphology of water droplets on silicone rubber surfaces exhibiting varying levels of hydrophobicity. The dissolution characteristics of soluble salts on contaminated silicone rubber were also analyzed in detail. The findings aim to provide a reference for the design and evaluation of artificial contamination tests for composite insulators.

7.2 Experimental Arrangement

7.2.1 Characterization for Wetting Degree of Hydrophobic Surfaces

Regardless of the impact of temperature differences, during the wetting process of artificial contamination, the water mist generated by the ultrasonic humidifier underwent constant collision and adsorption with the surface of silicone rubber. After the wetting process, as shown in Figure 7.1, two distinct regions formed on

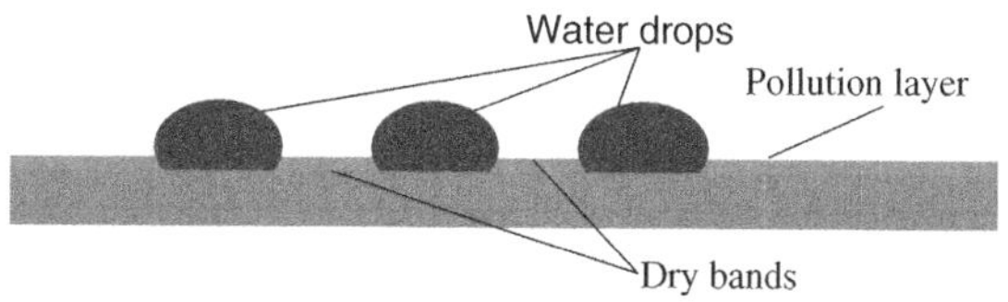

Figure 7.1 Description of surface wetting process. *Source*: Cao and Wang (2019)/with permission of Elsevier.

the silicone rubber surface: one consisting of water droplets, and the other composed of dry bands between the droplets. In the following, the dissolving characteristics of soluble salts on the two bands will be analyzed.

7.2.2 Test Specimens, Facilities, and Procedure

7.2.2.1 Test Specimens

The wetting process of artificial contamination and the dissolution process of soluble salts in the contamination were investigated. The silicone rubber specimens were vulcanized with hyperoxide in a mold at approximately 130 °C and 20 cm × 10 cm × 0.5 cm (length × width × thickness) in size. The main compositions of the silicone rubber used in the experiments are shown in Table 7.1 and the pigment and the vulcanizing agent in the silicone rubber are negligible. The silicone rubber material had good hydrophobicity and hydrophobicity transfer property.

7.2.2.2 Test Facilities

By controlling the exposure time of silicone rubber specimens in a controlled environment at 24 °C and 70% relative humidity, artificially contaminated samples with varying degrees of hydrophobicity were prepared. Considering that the static contact angle of contaminated silicone rubber specimens was affected by the surface roughness of contamination layers, which shows a certain dispersity, the surface hydrophobicity of test samples was divided into four hydrophobicity ranges: 25–30°, 55–60°, 75–80°, and 100–120°.

The wetting test platform used herein is shown in Figure 7.2. The test chamber had a size of 50 cm × 80 cm × 100 cm. The deionized water with temperature of 24 °C (same as ambient temperature in fog chamber) and conductivity of 2.4 mS/cm was transformed into micron water mist with ultrasonic humidifier and then accessed to the test chamber to provide moist environment. During the wetting process, the surface conductivity of specimens was recorded and measured by the high-precision RLC electric bridge. Because the voltage applied was low, the measuring current flowing through the contamination layer was very small and the thermal effect of current could be negligible. The specimens were placed horizontally in the test chamber and the contamination was never lost throughout the measuring process.

Table 7.1 Main compositions of the silicone rubber used in the experiments.

Components	Weight percentages (%)
Polymethylvinylsiloxane	34.48
Fumed silica	12.07
Aluminum hydroxide	53.45

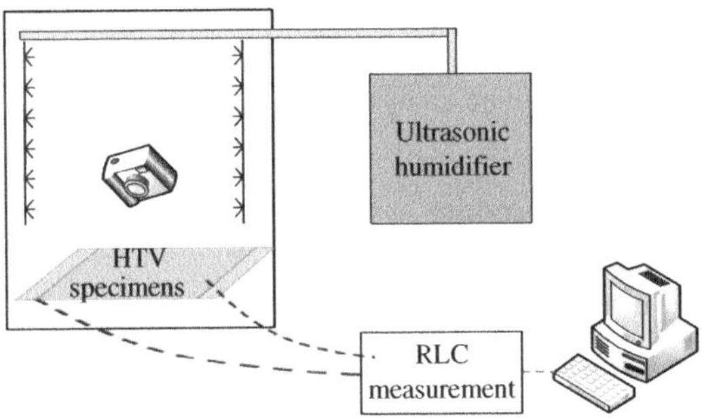

Figure 7.2 The wetting test platform. *Source*: Cao and Wang (2019)/with permission of Elsevier.

7.2.2.3 Maximum Wetting Time

Due to the discrete water drops instead of continuous water film formed on the hydrophobic surface after wetting, it was difficult to determine the saturated wetting point of the specimens directly from the aspects of leakage current or surface conductivity.

To estimate the wetting duration of each specimen, glass plates (with the same size and contamination levels: SDD = 0.1 mg/cm^2, NSDD = 1.0 mg/cm^2) were first tested in the fog chamber. During the wetting process, the surface conductivity of contamination layer was measured and Figure 7.3 presents the tendency of surface conductivity during the wetting process.

As illustrated in Figure 7.3, the wetting process comprises two stages. In the A–B stage, the contamination got wet gradually in the fog chamber and the soluble salts were dissolved progressively, which made the conductivity of contamination layer increase gradually. In the B–C stage, the soluble salts in the contamination layer dissolved completely as conductive ions. Although the amount of water attached to the glass surfaces continued to increase, the number of conductive ions in the layer already reached a maximum. Since the specimens were arranged horizontally, the contamination solution was not easily lost. Therefore, the surface conductivity of contamination layer no longer changed. At time point C, a thick layer of water film

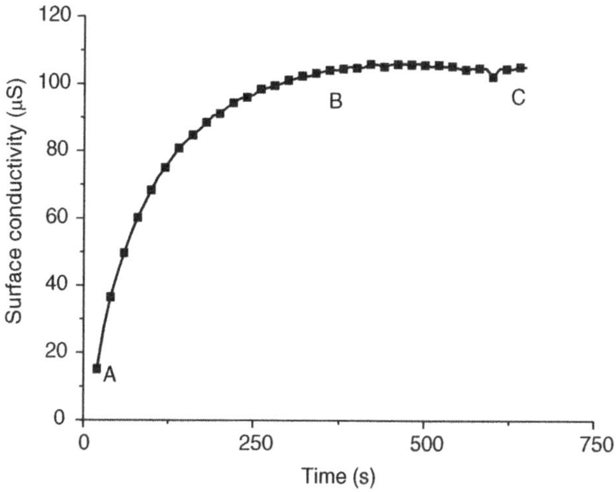

Figure 7.3 The tendency of surface conductivity during the wetting process. *Source:* Cao and Wang (2019)/with permission of Elsevier.

was observable on the glass surface and loss of contamination layer was visible by slightly tilting the specimen. Thus, the glass specimen can reach saturated moisture in only six minutes in the fog chamber designed herein. Considering that the surface of silicone rubber was hydrophobic, which was difficult to wet, the maximum time set for the tests herein was 16 minutes.

7.2.2.4 Test Procedure and Method

NaCl was used to simulate the soluble salt in contamination, whereas kaolin was used to simulate the non-soluble substances in contamination. To ensure the accurate contamination, the solid layer method was adopted herein to pollute the insulator surface. After each test, the contaminants were scrubbed off for measurement of SDD.

To accurately assess the quantity and salinity of water droplets formed on hydrophobic surfaces, four silicone rubber plates with identical contamination severity and hydrophobicity were prepared for each test group. The procedure was as follows:

1. Contaminate the surface of the silicone rubber specimens.
2. Place the specimens in a constant temperature and humidity chamber at 24 °C and 70% RH until the contamination layer reached a stable hydrophobic condition, which was quantified by contact angle measurements.
3. Place four specimens simultaneously in the fog chamber. Use the RLC bridge to measure the surface conductivity of one specimen.

4. Every four minutes, remove one specimen. Use dry absorbent paper to collect surface water droplets. Measure the mass of the absorbent paper before and after water absorption using a high-precision balance to determine total droplet mass.
5. Dissolve the used paper in 20 ml of water. Measure the solution salinity and calculate the salt concentration of droplets. The residual SDD on the contamination surface was also determined according to IEC 60815-1.

7.3 Results

7.3.1 Water Drop Salinity

The relationship between the surface water absorption of specimen and the wetting time is shown in Figure 7.4, while the dissolution ratios of soluble salts are shown in Figure 7.5.

As can be seen from Figure 7.4, the water absorption and the wetting time are basically linear since the specimens were all arranged horizontally. From Figure 7.5, it can be observed that over 60% of the soluble salts dissolved into water droplets within a short duration of wetting. With continued exposure, the amount of dissolved salt increased further, albeit at a much slower rate.

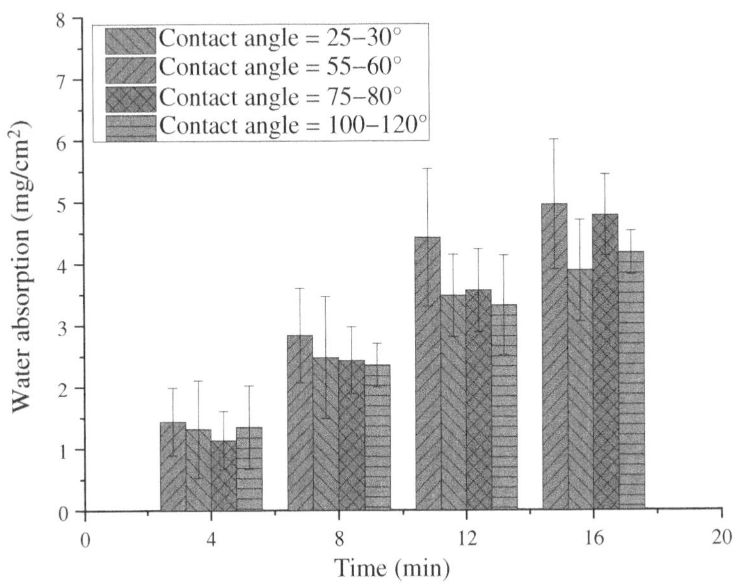

Figure 7.4 Variation of surface water absorption in the fog chamber.

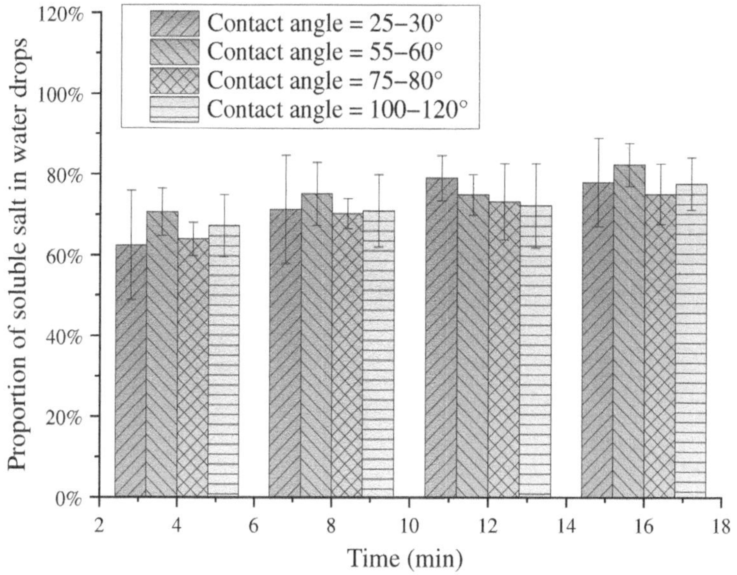

Figure 7.5 Proportion of soluble salts dissolved in water droplets on the HTV surface.

7.3.2 Surface Conductivity

During the wetting process, the resistances at both ends of the silicone rubber specimens were measured with RLC digital bridge. This was equivalent to measuring the resistance of circuit formed by parallel connection of bulk resistance and surface resistance. Since the bulk resistance of silicone rubber was in megohms, while the surface resistance after wetting was only a few thousand ohms, the measured resistance was basically the surface resistance of silicone rubber.

The surface conductivity K of the contamination layer was calculated using Eq. (7.1):

$$K = \frac{l}{R_d} \tag{7.1}$$

where l, d are the length and width of silicone rubber specimen, and R is the surface resistance measured by the digital bridge.

To normalize conductivity values to a standard temperature of 20 °C, the following correction was applied:

$$K_{20} = K\left[1 - b(T - 20)\right] \tag{7.2}$$

where b is the temperature coefficient, which can be obtained by table in IEC 60815-1, T is the actual ambient temperature during measurement.

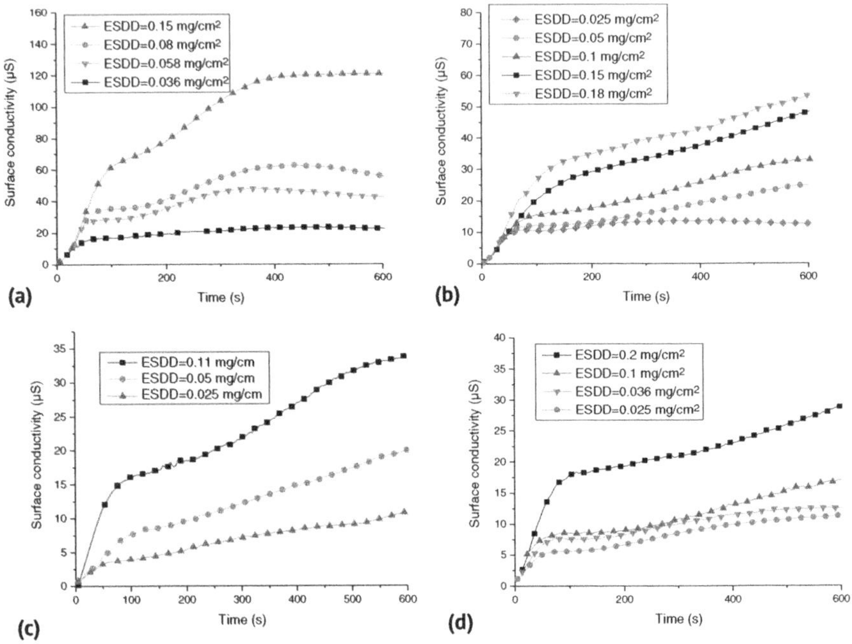

Figure 7.6 Variation of surface conductivity on hydrophobic surfaces during the wetting process: (a) contact angle = 25–30°; (b) contact angle = 55–60°; (c) contact angle = 75–80°; (d) contact angle = 100–120°. *Source*: Cao and Wang (2019)/with permission of Elsevier.

Figure 7.6 presents the variations of surface conductivity of specimens during the wetting process with varying contact angles. It is evident that the surface conductivity increased with wetting time, but this increase occurred in two distinct phases: In the initial stage, conductivity rose rapidly. After a certain point, the rate of increase significantly declined.

Despite an overall wetting time of 16 minutes, it can be seen from Figure 7.6 that after wetting for 10 minutes, the amount of water attached to the contamination layer already reached 4 mg/cm^2. This value had exceeded the amount water required to saturated absorption on the glass surfaces and visible water film would appear. However, it should be noted that within the actual wetting time of 10–16 minutes, the surface conductivity would still continue to rise.

The surface conductivity at 10 minutes in the wetting process was selected and plotted in Figure 7.7. As can be seen from Figure 7.7, the surface conductivity of specimens at 10 minutes is directly proportional to SDD. However, due to

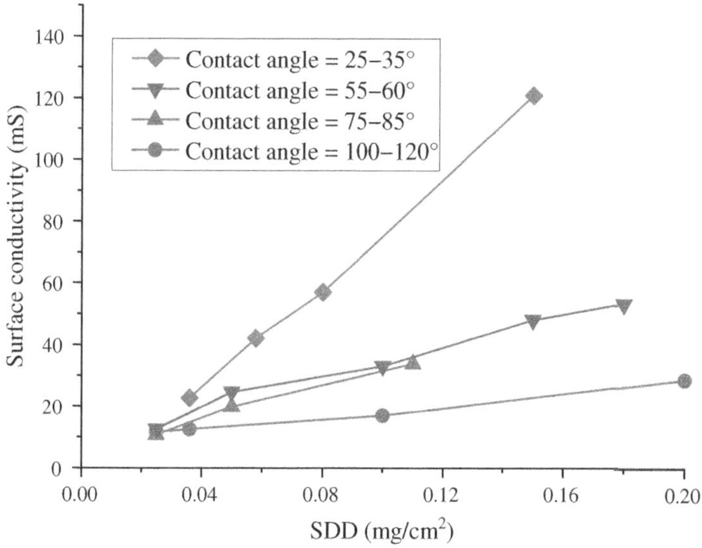

Figure 7.7 Relationship between surface conductivity at 10 minutes and SDD.

hydrophobic effects, specimens with higher hydrophobicity exhibited lower surface conductivity. This relationship can be described as:

$$K_{20} = p \times \text{SDD} \tag{7.3}$$

where p is related to the hydrophobicity of the contaminated silicone rubber surface. A higher degree of hydrophobicity corresponds to a smaller the p value.

7.4 Dissolving Characteristics of Soluble Salts at Water Drop and Dry Band

7.4.1 Dissolution Characteristics at Water Drop

The preceding analysis has shown that the soluble salts are greatly influential to the wetting process of hydrophobic surfaces. Therefore, more than 60% of salts were dissolved in water drops in a very short time in the wetting process. With the further prolongation of wetting time, the salt dissolved in water drops also increased further, but the rate of increase was slow. Thus, it can be concluded that part of salts was located in the area between water drops rather than in the location of water drops.

To further investigate this phenomenon, the wetting process of specimens was photographed and recorded with ANDONSTAR-A1 high-magnification microscope and the morphologies of water drops on the surface of contaminated silicone rubber with different hydrophobicity were observed. It can be found that in the wetting process, innumerable micron-sized water drops diffused in the fog chamber and finally adsorbed on the surface of silicone rubber. The entire wetting process is shown in Figure 7.8.

NaCl is used as a substitute for the soluble salt in the contamination in the artificial pollution test. During the drying process the sodium chloride originally mixed evenly in the contamination layer gradually recrystallized and precipitated out of the layer and it had an obvious impact on the wetting of contamination layer. Microscopic observations revealed that wetting did not occur uniformly across the surface. Instead, water droplets preferentially appeared at salt recrystallization zones, as shown in Figure 7.9.

When the surface was coated solely with kaolin and no soluble salts, the randomness of water droplet locations during wetting increased significantly. In this case, no distinct wetting pattern was observed. For contamination layers with low hydrophobicity, airborne water mist particles collided with the surface and

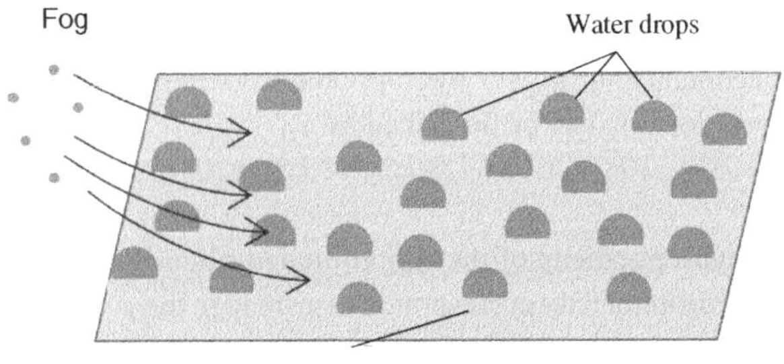

Figure 7.8 Wetting process of contaminated silicone rubber surface. *Source*: Cao and Wang (2019)/with permission of Elsevier.

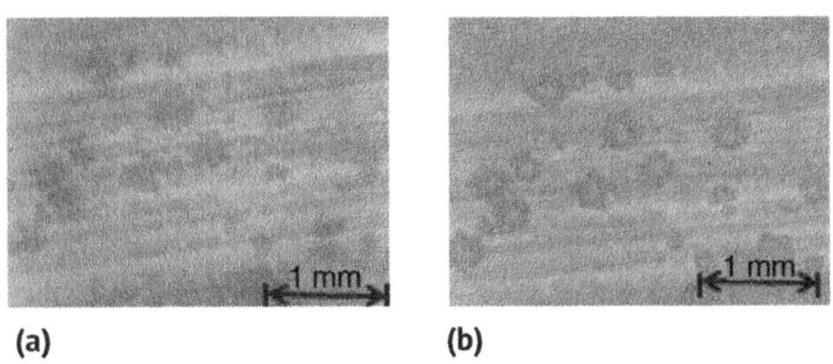

Figure 7.9 Effect of NaCl on the wetting of contamination layers: (a) before wetting; (b) after wetting. *Source*: Cao and Wang (2019)/with permission of Elsevier.

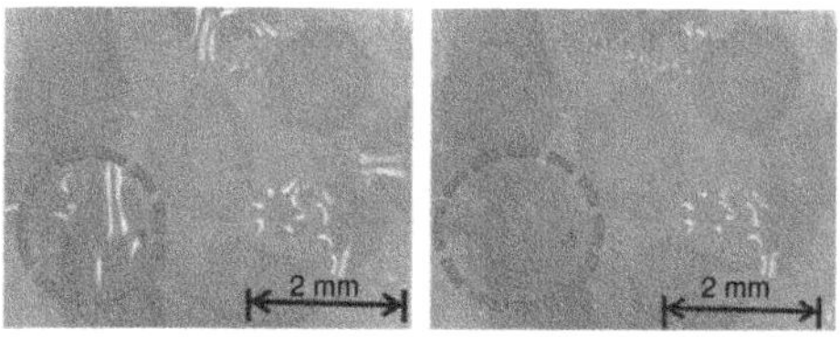

(a) (b)

Figure 7.10 Water droplet coalescence on HTV surface (contact angle = 66.3°): (a) before merging; (b) after merging. *Source*: Cao and Wang (2019)/with permission of Elsevier.

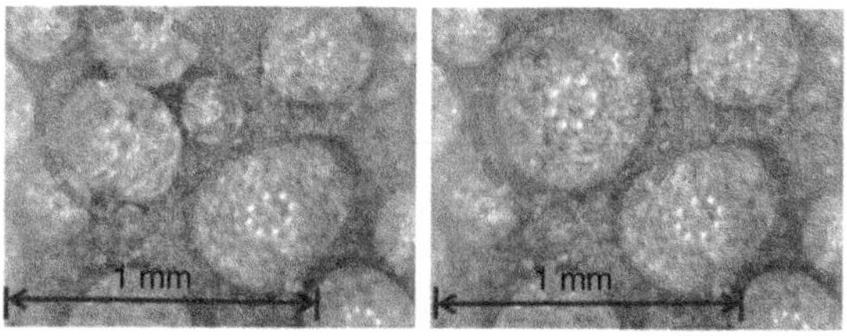

(a) (b)

Figure 7.11 Water droplet coalescence and relocation on HTV surface (contact angle = 117.3°): (a) before merging; (b) after merging. *Source*: Cao and Wang (2019)/with permission of Elsevier.

quickly coalesced into larger droplets. As these droplets grew, they merged in place, increasing in size but not shifting position, as shown in Figure 7.10.

However, for highly hydrophobic silicone rubber surfaces, droplets exhibited both growth and movement. Upon reaching a critical size, droplets not only merged but also shifted location, as shown in Figure 7.11.

In artificially contaminated specimens, the salt recrystallization regions were consistently the initial wetting sites, with subsequent expansion of wetted areas. This indicates that the soluble salts are greatly influential to the wetting process of hydrophobic surfaces. However, the process of natural contamination is much more different and the detailed analysis will be presented in another article.

7.4.2 Dissolving Characteristics of Soluble Salts at the Dry Bands

For the dry bands on the artificial contaminated silicone rubber, the salt dissolution rate during wetting can be characterized by the surface conductivity. This was because the larger the dissolved amount of soluble salts in contamination is, the greater the number of electrically conductive ions in the contamination is, and the higher the conductivity of contamination layer is. As for the hydrophilic surfaces, the variation trend of surface conductivity over wetting time is shown in Figure 7.12.

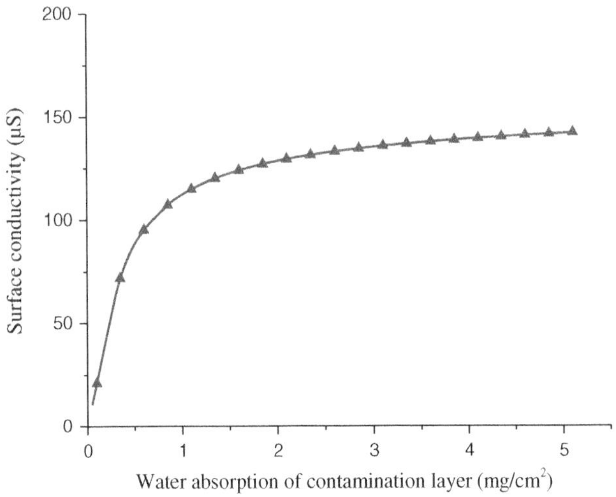

Figure 7.12 Theoretical relationship between surface conductivity and water absorption (SDD = 0.1 mg/cm^2).

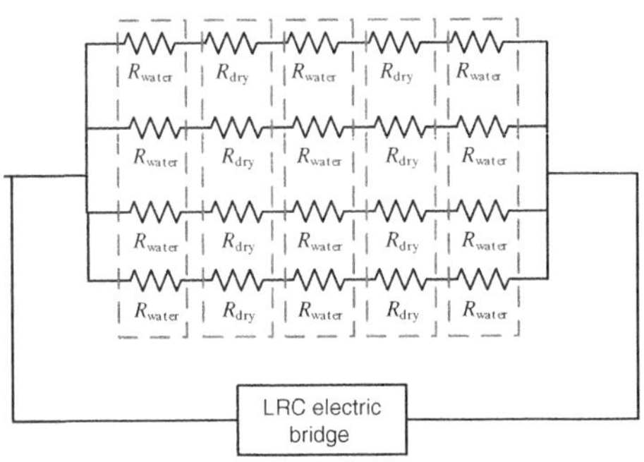

Figure 7.13 Equivalent electrical circuit of the wetting contamination layer. *Source*: Cao and Wang (2019)/with permission of Elsevier.

To assess the dissolved salt content in dry bands between droplets, one only needs to measure the surface conductivity at the dry bands. However, in practical wetting scenarios, dry bands and water droplets form a series circuit, as shown in Figure 7.13. Thus, the total measured resistance represents the equivalent resistance of the entire surface:

$$R_{\text{test}}(t) \approx R_{\text{dry}}(t) + R_{\text{water}}(t) \tag{7.4}$$

where $R_{\text{test}}(t)$ is the overall equivalent resistance during wetting, $R_{\text{dry}}(t)$ is the equivalent surface resistance of contamination layer at the dry bands, and $R_{\text{water}}(t)$ is the equivalent surface resistance of water drops. Therefore, the relationship between dry band surface conductivity and overall surface conductivity still requires further analysis.

The above results on the salinity of water drops already demonstrate that during the wetting process, the water drops are always formed at the salt recrystallization area on the silicone rubber surface first. The salinity and conductivity of the water drops is high. Since these droplets are connected in series with the dry bands, the measured surface resistance mainly reflects the resistance of the dry bands:

$$R_{\text{test}} \approx R_{\text{dry}}(t) \quad 0 < t < t_0 \tag{7.5}$$

where t_0 is the moment before which the resistance of water drops is far less than the resistance of dry bands.

Therefore, the measured conductivity also reflected the dissolution of soluble salts between water drops, which can be expressed as:

$$K_{\text{dry}}(t) \approx K_{\text{test}}(t) \quad 0 < t < t_0 \tag{7.6}$$

As wetting progresses and water droplets expand, dry band areas narrow. The high salinity in water droplets causes a decrease in overall resistance, and the surface conductivity measurement no longer accurately reflects the dry band salt dissolution.

Considering that the test specimens were arranged horizontally during wetting in the fog chamber, and that the flow rate of mist was unchanged throughout the tests, the moisture absorptions of test specimens per unit time were the same in each set of tests. For the artificially contaminated silicone rubber specimens with the same SDD and different hydrophobicity, the variation trends of surface conductivity in the early wetting time are as shown in Figure 7.14.

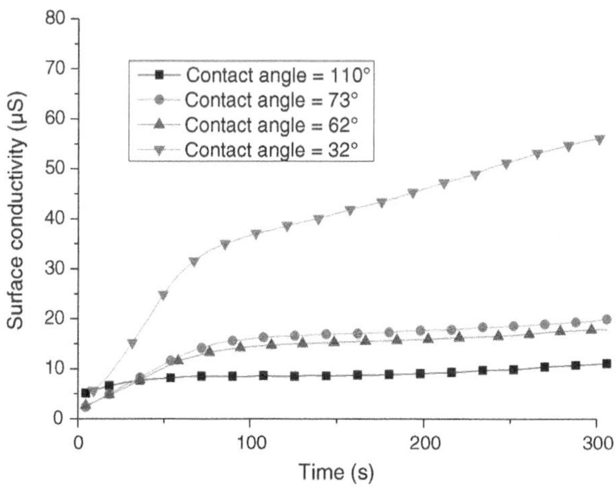

Figure 7.14 Surface conductivity trends during early wetting phase. *Source:* Cao and Wang (2019)/with permission of Elsevier.

As can be seen in Figure 7.14, for the contaminated silicone rubber specimens in the early stage of wetting, the variation of surface conductivity can be roughly divided into two stages depending on the rate of change. In the first stage, the surface conductivity increased rapidly, whereas in the second stage, the surface conductivity grew slowly. Detailed analysis revealed that specimens with lower hydrophobicity exhibited a faster increase in conductivity during the initial wetting stage, indicating a quicker dissolution of soluble salts in the dry bands between water droplets.

7.5 Summary

This chapter investigated the wetting behavior and salt dissolution characteristics of artificially contaminated silicone rubber surfaces. The following conclusions were drawn:

1. For the artificial contaminated silicone rubber, the water droplets preferentially formed at the salt recrystallization zones due to the hygroscopic nature of NaCl. This resulted in water droplets with relatively high salinity and conductivity, regardless of the surface hydrophobicity.
2. Throughout the wetting period, although a continuous water film does not form between the droplets, the surface conductivity of the contamination still increases. This indicates that the soluble salts in the contamination layer of the dry band also partially dissolve to form conductive particles.
3. The amount of soluble salts in the dry bands is influenced by the surface hydrophobicity of HTV silicone rubber specimens. The stronger the hydrophobicity of contaminated silicone rubber surface, the less the dissolved amount of soluble salts.

Future work will focus on investigating the wetting behavior of silicone rubber contaminated with various types of soluble salts. Additionally, the correlation and equivalency between natural and artificial contamination will be further examined to enhance the representativeness of laboratory tests.

Bibliography

Abbasi, A., Shayegani, A., and Niayesh, K. Contribution of design parameters of SiR insulators to their DC pollution flashover performance, IEEE Trans. Power Del., 2014, 29 (4), 1814–1821.

Arias Velásquez, R.M. and Mejía Lara, J.V. Current transformer failure caused by electric field associated to circuit breaker and pollution in 500 kV substations, Eng. Fail. Anal., 2018, 92, 163–181.

Arias Velásquez, R.M. and Mejia Lara, J.V. The need of creating a new nominal creepage distance in accordance with heaviest pollution 500kV overhead line insulators, Eng. Fail. Anal., 2018, 86, 21–32.

Cao, B. and Wang, L. Wetting characteristics of artificial contamination on the hydrophobic surface, Eng. Fail. Anal., 2019, 100, 428–438.

Chrzan, K. Conductivity of aqueous dust solutions, IEEE Trans. Electr. Insul., 1987, EI-22 (3), 241–244.

Dong, B., Jiang, X., Zhang, Z. et al. Effect of environment factors on ac flashover performance of 3 units of polluted insulator strings under natural fog condition, IEEE Trans. Dielectr. Electr. Insul., 2014, 21 (4), 1926–1932.

Du, B.X. and Liu, Y. Pattern analysis of discharge characteristics for hydrophobicity evaluation of polymer insulator, IEEE Trans. Dielectr. Electr. Insul., 2011, 18 (1), 114–121.

Farzaneh, M. and Chisholm, W.A. Insulators for Icing and Polluted Environments, John Wiley & Sons, 2009.

Hanqi, D., Hongwei, M., and Xilin, W. Influence of air flow on hydrophobicity transfer of polluted high temperature vulcanized silicone rubber samples, High Volt. Eng., 2015, 41 (8), 2795–2802.

Ilhan, S. and Cherney, E.A. Comparative tests on RTV silicone rubber coated porcelain suspension insulators in a salt-fog chamber, IEEE Trans. Dielectr. Electr. Insul., 2018, 25 (3), 947–953.

International Electrotechnical Commission (IEC) 60815-1 2008-10, Selection and Dimensioning of High-voltage Intended for Use in Polluted Conditions – Part 1 Definitions, Information and General Principles, 2008.

Jiang, X., Yuan, J., Zhang, Z. et al. Study on AC pollution flashover performance of composite insulators at high altitude sites of 2800-4500 m, IEEE Trans. Dielectr. Electr. Insul., 2009, 16 (1), 123–132.

Shah, M., Karady, G.G., and Brown, R.L. Flashover mechanism of silicone-rubber insulators used for outdoor insulation-2, IEEE Trans. Power Del., 1995, 10 (4), 1972–1978.

Verma, A.R. and Reddy, B.S. Interpretation of surface degradation on polymeric insulators, Eng. Fail. Anal., 2019, 95, 214–225.

Verma, A.R. and Reddy, B.S. Tracking and erosion resistance of LSR and HTV silicon rubber samples under acid rain conditions, IEEE Trans. Dielectr. Electr. Insul., 2018, 25 (1), 46–52.

Wang, L., Cao, B., Mei, H. et al. Effects of natural contamination components on the surface conductivity under saturated moisture, IEEE Trans. Dielectr. Electr. Insul., 2017, 24 (5), 2945–2951.

Xidong, L., Chao, W., Yiming, Y. et al. Effective equivalent salt deposit density of polluted silicone rubber insulators in wetting process, 2014 IEEE Conference on Electrical Insulation and Dielectric Phenomena (CEIDP), 2014, pp. 271–274.

Xidong, L., Yanfeng, G., Yibo, Z. et al. Modeling of the salt dissolving process of wet natural pollution on silicone rubber insulators, 2013 Annual Report Conference on Electrical Insulation and Dielectric Phenomena, 2013, pp. 444–447.

Yang, L., Zhang, F., Wang, Z.H. et al. Evaluation of wetting condition and its effects on pollution flashover voltage of aerodynamic insulators, IEEE Trans. Dielectr. Electr. Insul., 2016, 23 (5), 2875–2882.

Zhang, C. and Wang, L. Experimental investigation on pollution flashover performance of multiple parallel suspension insulators, IEEE Trans. Dielectr. Electr. Insul., 2016, 23 (5), 2840–2849.

Zhang, C., Wang, L., Guan, Z. et al. Pollution flashover performance of full-scale ±800 kV converter station post insulators at high altitude area, IEEE Trans. Dielectr. Electr. Insul., 2013, 20 (3), 717–726.

Zhang, F., Wang, L., Guan, Z. et al. Influence of composite insulator shed design on contamination flashover performance at high altitudes, IEEE Trans. Dielectr. Electr. Insul., 2011, 18 (3), 739–744.

Zhang, F., Wang, X., Wang, L. et al. Effect of arcing on DC flashover performance of contaminated porcelain insulators for various suspension patterns at high altitudes, IEEE Trans. Dielectr. Electr. Insul., 2008, 15 (3), 783–791.

8

Wetting Characteristics of the Hydrophobic Material Surface Through Condensation

8.1 Introduction

As discussed in Chapters 2 and 3, the presence of water significantly influences the occurrence of flashover, whether it involves surface flashover or air breakdown. Water droplets on the surface can alter the local electric field distribution and effectively reduce the insulation distance.

On hydrophobic surface, the flashover mechanism is quite different from that on hydrophilic surfaces, primarily due to the isolated nature of water droplets. The water droplets may cause flashover with fast speed and unobvious leakage current. Consequently, for highly hydrophobic insulators, the primary concern shifts from salt or pollution density to the formation of water droplets.

It is well established that both temperature and humidity have a great influence on the condensation process. However, in the context of insulator operation, the high-voltage factor should be considered as well. When the air is below the dew point due to the radiative cooling, the saturated water vapor will condense to water droplets on the surface of the insulator.

Condensation can moisten the insulator surface without necessarily removing the existing pollution layer. These accidents in the desert area, indoor flashover accidents, and dawn trip accidents are all caused by the condensation on the insulator. Based on that, the wetting process on the insulator or hydrophobic surface is the key point in preventing these flashover accidents.

To this end, this chapter investigates the influence of temperature differentials on the wetting process of hydrophobic materials. Specifically, it examines variations in the mass, length, and surface area of water droplets during the wetting process under temperature gradients (hereafter referring to the temperature difference between ambient air and hydrophobic surfaces). Furthermore, the relationship between wetting and time cost for flashover is also discussed.

8.2 Experimental Arrangement

8.2.1 Test Arrangement and Test Circuit

The test platform must be capable of controlling both temperature and electric field to simulate condensation conditions. To study the wetting characteristics caused by condensation of the hydrophobic material, a test platform was built as shown in Figure 8.1.

The temperature control system consists of a semiconductor cooling module, a cooling pad, a power supply (12 V lead-acid battery), a thin-film temperature sensor, a thermostat, and cooling fans. The cooling pad is positioned beneath the test specimen to facilitate efficient heat transfer. A thermostat employing PID (Proportional-Integral-Derivative) control ensures precise temperature regulation of the specimen surface, maintaining fluctuations within ±1 °C. Both the cooling components and thermostat are powered by a 12 V lead-acid battery.

To generate the required high electric field, a 50 kV/100 kVA test transformer is employed. Meanwhile, the same data acquisition (DAQ) system in Chapter 2 is used to measure the voltage signal during the wetting experiment.

8.2.2 The Design and Selection of Electrodes

The electrodes used on the specimen need to be simulated and tested. Based on experiments requirements, two rod electrodes were used and the placement of electrodes is shown in Figure 8.2. The rod electrodes are 5.0 cm in length and 0.5 cm in width.

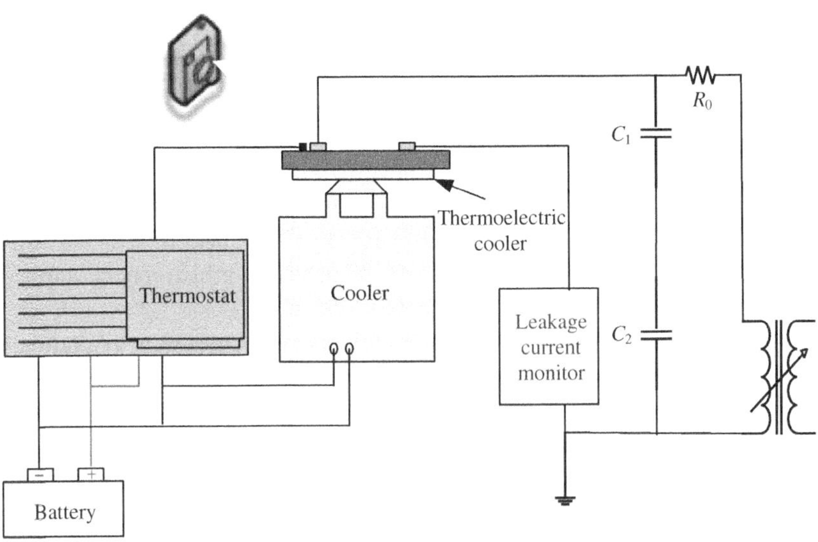

Figure 8.1 Schematic of the temperature control system and high-voltage test circuit. *Source*: Wu et al. (2023)/John Wiley & Sons/CC BY 4.0.

A 1.0 cm^2 square zone is marked as the sampling area. To avoid the distortion of the electric field, the four corners of the electrodes are rounded and polished. In order to test the electrodes, the electrodes were placed on an HTV silicone rubber plate ($D = 15.0$ cm) with 90°~93° contact angle. A preliminary test with 15 °C temperature between the air and hydrophobic surface is carried out as Figure 8.3 shows.

As shown in Figure 8.3, water droplet distribution is nonuniform. At both 12 and 18 minutes, the condensation is mainly concentrated around the high-voltage (HV) electrode, which is highlighted with the half circle. It is speculated that the uneven distribution of water droplets is caused by the electrical field so the simulation of the electric field has been done in Figure 8.4.

The electric field is intensified and uneven around the end of the HV electrode. Meanwhile, the round shape distortion region has explained why the condensation is inside the half circle in Figure 8.3. Through the preliminary test, it is known that the electric field has a certain effect on the wetting process. Therefore, a relatively uniform electric field is required. Besides this, to study the wetting process, multiple sampling areas are needed as well. After reconsideration, the electrodes are designed as Figure 8.5.

Figure 8.2 Rod electrodes.

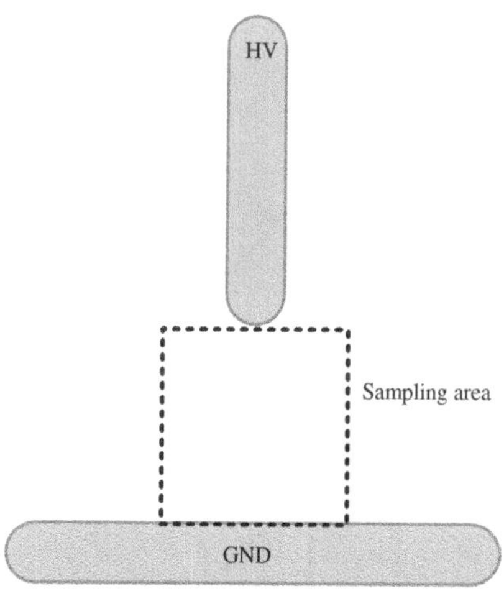

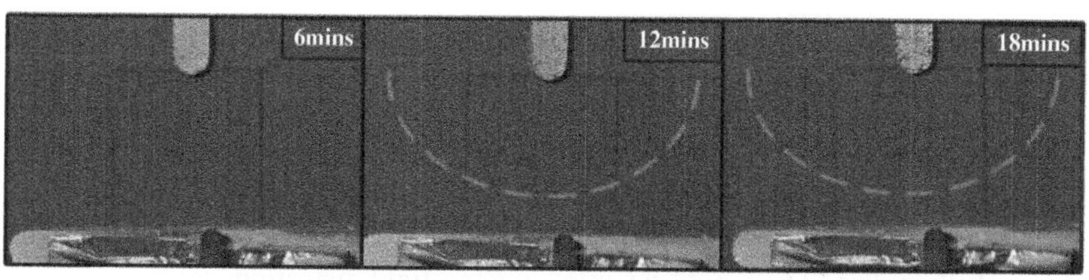

Figure 8.3 Preliminary test.

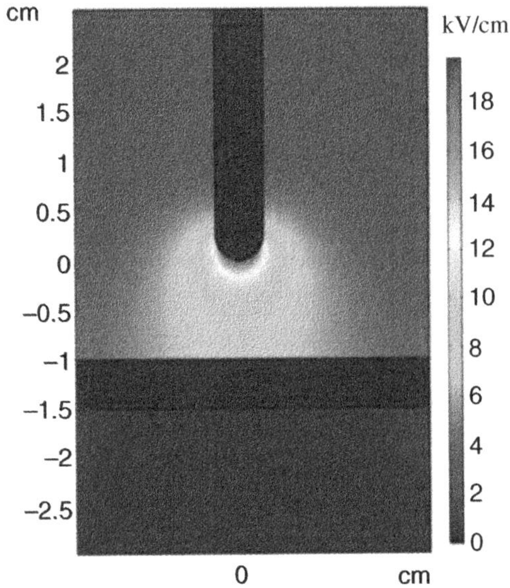

Figure 8.4 Simulation result of the preliminary test.

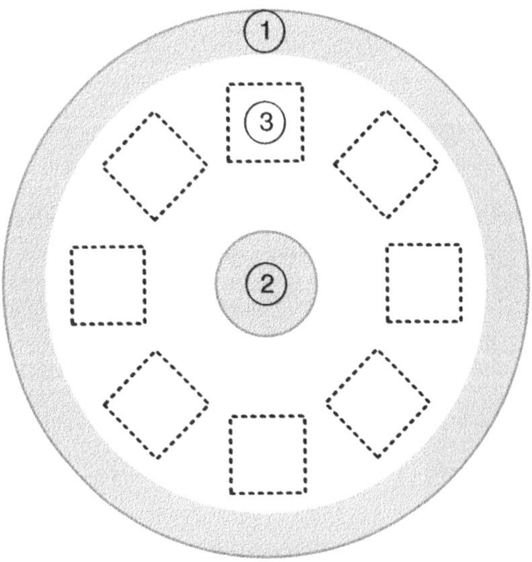

Figure 8.5 Electrodes design. 1: HV electrode 2: Ground electrode 3: Sampling area.

The redesigned electrode system consists of a circular HV electrode with inner and outer radii of 3.0 and 3.5 cm, respectively, and a single circular ground electrode with a radius of 1.0 cm. Both of the electrodes are 1.0 mm thick copper and are carefully polished to avoid unwanted electric field distortion.

The electrodes are placed on an HTV silicon rubber disk with a cooler on its backside. To enhance thermal conduction, the HTV specimen must be thin. However, the cooler on the backside is made of metal which may distort the electric field on the specimen's surface. The specimen needs to be thick enough to limit the electric field distortion caused by the cooling pad. After evaluation, a 6 mm thick HTV silicone rubber plate was chosen. To verify the electric field distribution on its surface, the electric field with and without a cooler on backside were simulated.

The applied voltage is 6 kV, and the maximum electric field in these two cases are both around 8 kV/cm at the edge of the ground electrode. Comparing with the electric field in the preliminary test, the electric field in the test area is much lower and uniform, as shown in Figure 8.6. And the electric field difference caused by the cooler on the backside is neglectable. Thus, this pair of circular copper electrodes was selected for subsequent experiments.

8.2.3 Specimen Arrangement

The specimen arrangement is shown in Figure 8.7. During the wetting process, the temperature and relative humidity of the environment were constant and the temperature of test specimens was controlled by a semiconductor cooler which was pasted on the back of the specimen. So as to accurately control the temperature difference, a thin temperature sensor was fixed on the HTV surface. PID control method was adopted to keep the temperature of the specimen surface constant with 1 °C deviation.

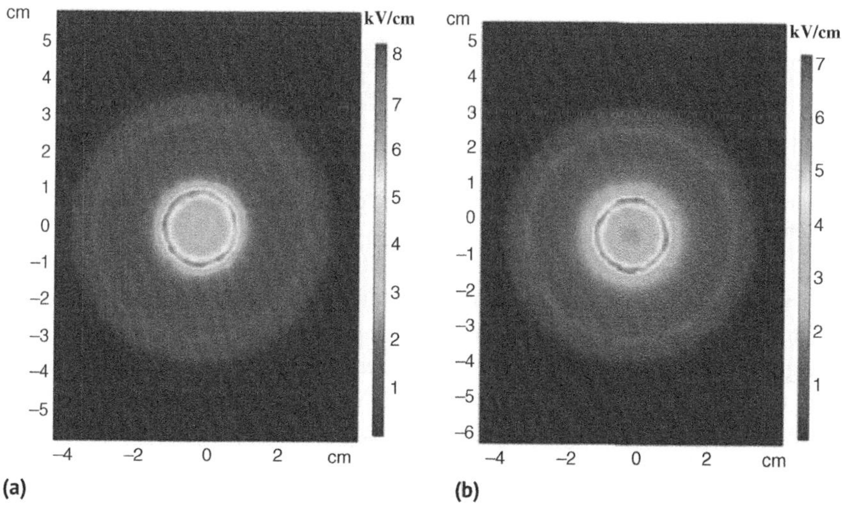

Figure 8.6 Electric field comparison with and without cooler on the backside: (a) with cooler; (b) without cooler.

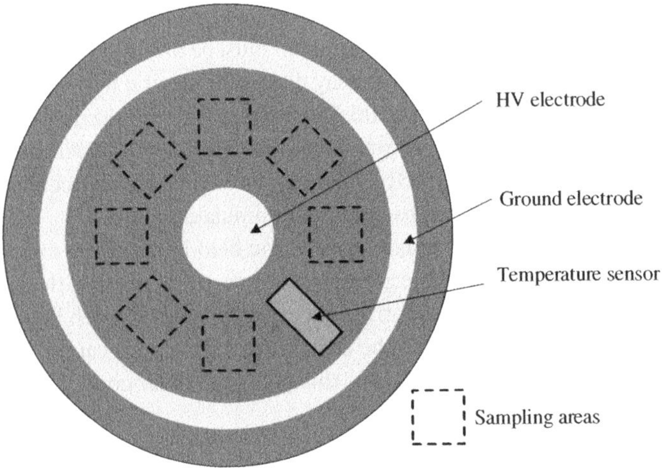

Figure 8.7 Specimen arrangement. *Source*: Wu et al. (2023)/John Wiley & Sons/CC BY 4.0.

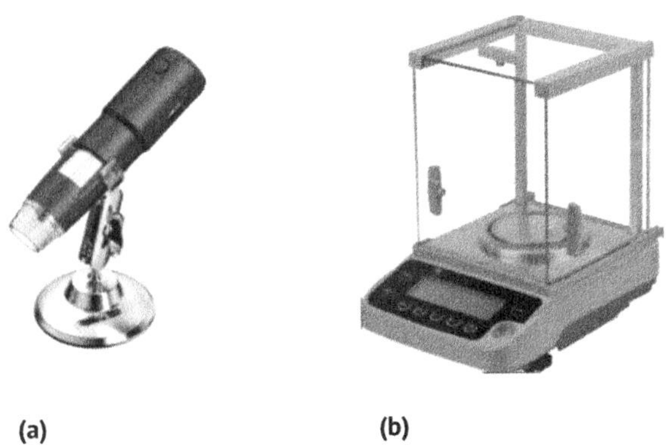

Figure 8.8 Measuring equipment for the wetting process: (a) microscope; (b) electronic balance.

The mass of condensation was also measured to describe the wetting rate. In each experiment, water droplets in the seven sampling areas were collected and weighed by using absorbent paper every three minutes in a clockwise sequence and the water droplet morphology on the HTV surface was recorded by a microscope (Leipan MS5). The electronic balance and microscope used in experiments are shown in Figure 8.8.

Considering that the time used for the initial cooling might be different in each experiment, before the surface temperature cooled down to the set value, the sample is wrapped in plastic wrap to prevent wetting.

In this section, the effect of electric field on the wetting of test specimens with electric field were studied by the control variate method. The temperature difference was controlled at 10 °C, 13 °C, and 16 °C, and the electric field on the test specimens was set to 0 kV/cm, 1.0 kV/cm, 2.0 kV/cm, 2.5 kV/cm and 3.0 kV/cm, respectively, by adjusting the applied voltage.

The previous research in Chapters 2 and 3 showed the flashover voltage of a hydrophobic surface with the same pollution degree is related to the water proportion along the insulation length after wetting. To find out the relationship between withstanding time and different temperature differences, the water droplet area proportion k_A and water droplet insulation length proportion k_L during the wetting process will be characterized and analyzed in the next section.

8.2.4 The Image Processing for Water Proportion

As it has been discussed in previous chapters, the proportion of water droplets to insulation length is vital to the occurrence of flashover, which is determined by the residual insulation length. So, it is worthy to study the water proportion during the wetting process. To quantify this, two parameters are proposed to characterize the degree of wetting on the hydrophobic surface:

1. k_A: The area proportion of water droplets on the test specimens during the wetting process.
2. k_L: The lengthwise proportion of water droplets along the insulation path, representing the fraction of the insulation length that is covered by droplets.

The water proportion was determined using an image processing algorithm developed in MATLAB. The procedure is outlined below, and the corresponding workflow is shown in Figure 8.9:

1. Select the same sampling area as the analysis object in wetting photos every three minutes.
2. Preprocess the images by adjusting contrast and grayscale levels to ensure consistency across different frames.
3. Select the proper threshold value and then binarize the preprocessed photo.
4. Calculate the proportion of black pixels to total pixels, this result refers to the area proportion of water droplets which is k_A.

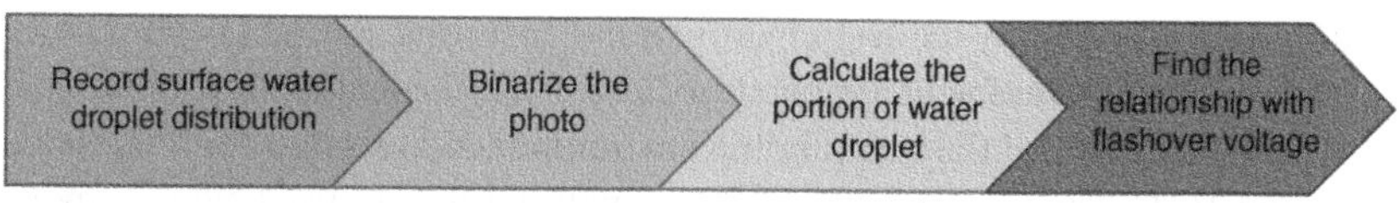

Figure 8.9 The image processing procedure.

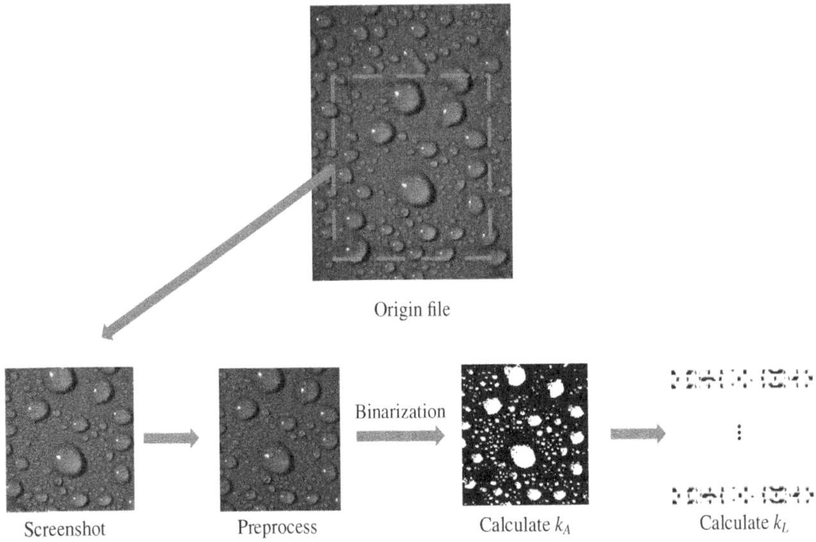

Figure 8.10 Actual image processing example.

5. Recalculate from step three, equally divide the binary images into 100 equal segments along the horizon direction.
6. Calculate each proportion of black pixels to total pixels of 100 parts, the maximum result refers to the length proportion of water droplet between two electrodes which is k_L.

Figure 8.10 represents a typical image processing procedure for k_A and k_L. By means of above the image process method, the area proportion and length proportion of water droplets can be calculated. All the photos in experiments have been analyzed to compare the water droplet proportion during the wetting process.

8.3 Results

8.3.1 Effect of Initial Temperature Difference

According to Eq. (5.10) in Chapter 5, the dew point temperature under this condition was 19.5 °C in the test environment (28 °C and RH 60%), which was 8.5 °C lower than the ambient temperature. Therefore, the temperature difference between specimens and the surrounding environment was set as 5 °C, 7 °C, 10 °C, 13 °C, and 16 °C. The typical wetting process without the electric field is demonstrated in Figure 8.11. In the early stages—prior to 12 minutes—small and densely packed water droplets rapidly formed on the specimen surface, with the number of visible droplets increasing over time. In the final three frames, the droplets began to coalesce into larger clusters. While in the last three pictures, the water droplets started to merge. The distribution of water droplets during the wetting process was quite uniform. The evolution of water mass during the wetting process under temperature difference was shown in Figure 8.12.

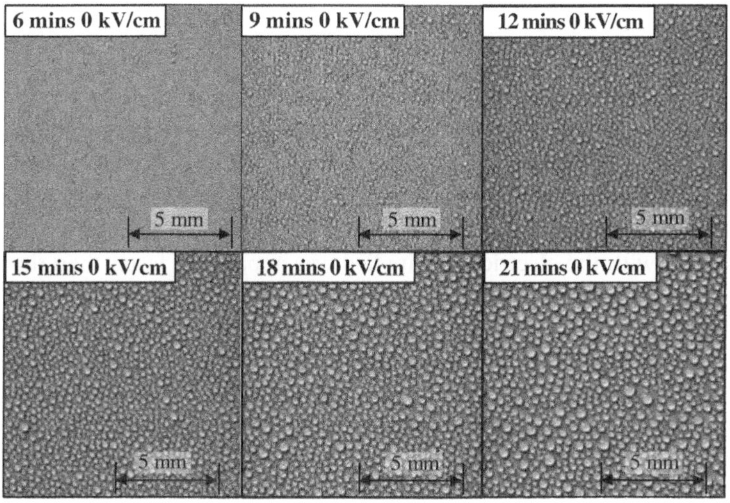

Figure 8.11 Typical wetting process at different time points without electric field ($T = 28\,°C$, $\Delta T = 13\,°C$, RH = 78%).

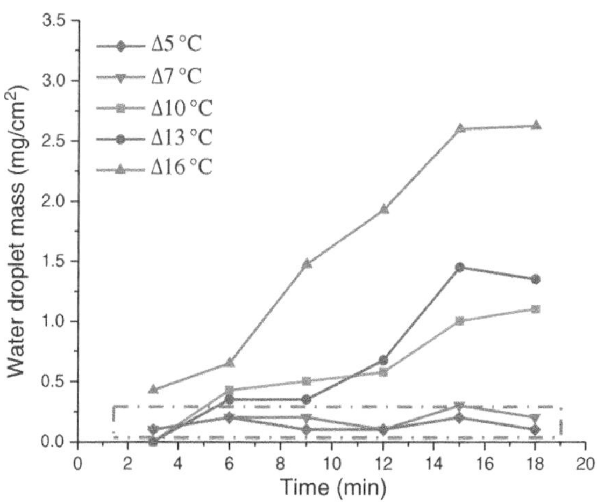

Figure 8.12 Relationship between water droplet mass and time without electric field ($T = 28\,°C$, RH = 60%).

When the specimen temperature is higher than the dew point temperature (7 °C and 5 °C), there was no water formed on the specimens' surface during the whole test. However, once the dew point is satisfied the visible water droplets appeared at six minutes, and the water mass also started to increase. With a higher temperature difference, the wetting rate was higher and the water mass at the end of the test (after 21 minutes) was higher as well.

8.3.2 Effect of Electric Field

The evolution of water droplet mass during the wetting process with the 1.5 kV/cm electric field was shown in Figure 8.13, with 28 °C room temperature and 76% relative humidity. As the same phenomenon as the experiment without an electric field, water droplets only occurred when the surface temperature of the test specimens was below the dew point and it could also observe that the wetting rate with the electric field applied was apparently faster than that without the electric field.

The morphology of water droplets during the wetting process with the electric field was shown in Figure 8.14. Compared with the test result without an applied voltage, field distortion induced by the electric field caused small droplets to coalesce into larger ones more quickly. At 21 minutes, the droplet size was noticeably larger and the distribution across the surface more uneven.

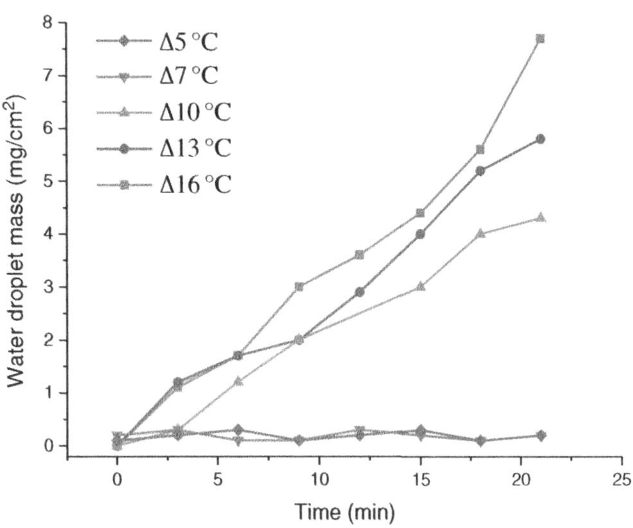

Figure 8.13 Relationship between water droplet mass and time with the electric field ($T = 28\,°C$, RH = 76%).

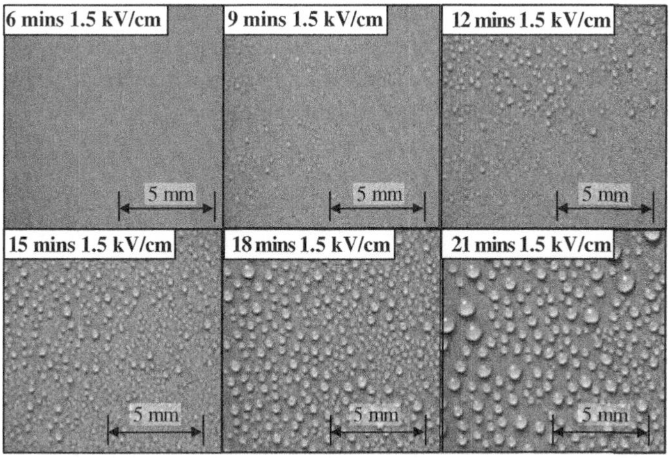

Figure 8.14 Typical wetting process at different time points with the electric field ($T = 28\,°C$, $\Delta T = 13\,°C$, RH = 76%). *Source*: Wu et al. (2023)/John Wiley & Sons/CC BY 4.0.

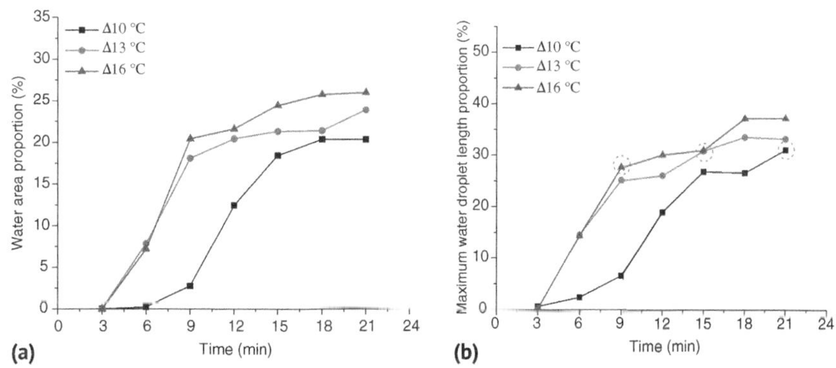

Figure 8.15 Water droplet proportion k_A and k_L versus time: (a) water droplet area proportion k_A; (b) water droplet coverage along insulation length k_L. *Source*: Wu et al. (2023)/John Wiley & Sons/CC BY 4.0.

To quantitatively analyze the effect of the electric field on the wetting process, images of the specimen surface were binarized to extract the water droplet area proportion (k_A) and the insulation length coverage proportion (k_L), as shown in Figure 8.15. When the temperature difference increases, the rising speed and final value of k_A and k_L both rise accordingly. The water droplet with a higher temperature difference has relatively larger diameter and covers more hydrophobic areas.

The final water droplet area proportion of 21 minutes of wetting under these three different temperature differences are 20.3, 23.9, and 26.0% respectively. This is because under AC electric field the water droplet oscillates by electric force and the vibration of water droplets will lead to the collapse and merge of the water droplet. Under the influence of electric force, these small water droplets will merge to larger size of water droplets which will make k_A and k_L both increases.

8.4 Discussion

8.4.1 Electric Field Strength

In this experiment, the temperature difference between the surrounding air and the test specimens was maintained at 13 °C, and the air temperature and relative humidity were 28 °C and 75%. The experiment variable is electric field strength. By applying different AC voltage, 0 kV/cm, 1.0 kV/cm, 2.0 kV/cm, 2.5 kV/cm, and 3.0 kV/cm electric fields would form at the sample zone respectively, the condensation mass and surface morphology were recorded at the same time. The wetting process with different electric fields is demonstrated below.

As shown in Figure 8.16, the morphology of water droplets—including size, shape, and spatial distribution—was significantly affected by the electric field

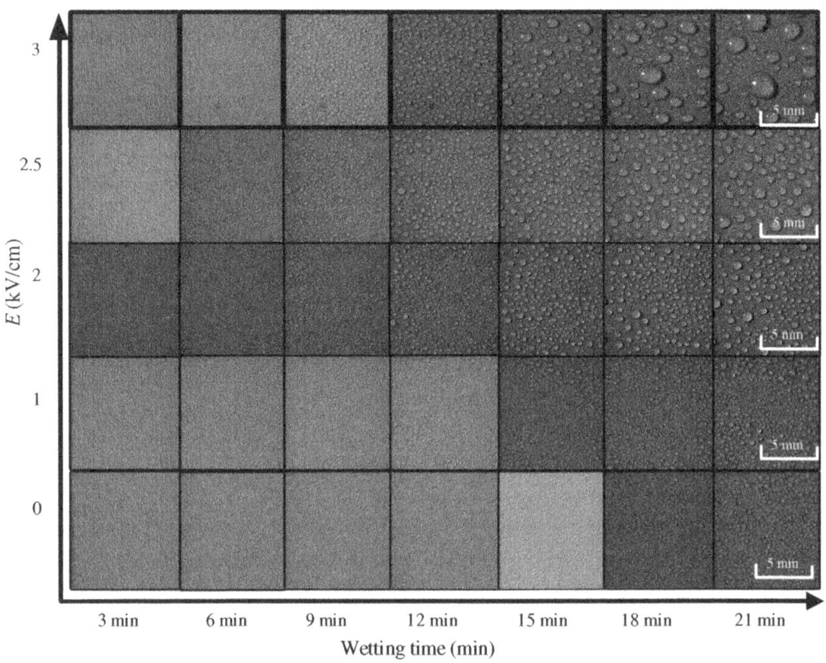

Figure 8.16 Wetting process at different time points with different electric field strengths ($T = 28$ °C, $\Delta T = 13$ °C, RH = 75%). *Source:* Wu et al. (2023)/John Wiley & Sons/CC BY 4.0.

strength. The higher electric field will make the moment for visible water droplets to appear earlier and the largest droplet at 21 minutes is larger as well. However, with an electric field, the water droplets were merged by electric force and generated large size droplets, the size and distribution of droplets were irregular.

The water droplet mass versus time t with different field strengths is presented in Figure 8.17. As electric field strength increased, both the rate of condensation and the final water mass also increased. Notably, under a fixed electric field, the water accumulation rate accelerated over time. This can be attributed to the influence of preexisting water droplets, which further distorted the local electric field, enhancing subsequent condensation.

Water droplet mass $m_{2.5}$ with 2.5 kV/cm and $m_{3.0}$ with 3.0 kV/cm increase with an exponential trend, a_0 and b_0 are constants and R_1, R_2 are the correlation coefficients. The fitting equations are as follows:

$$\begin{cases} m_{2.5} = \mathrm{a}_0 + 0.00929e^{0.03321t} \\ {R_1}^2 = 0.99824 \end{cases} \tag{8.1}$$

$$\begin{cases} m_{3.0} = \mathrm{b}_0 + 0.00465e^{0.05653t} \\ {R_2}^2 = 0.99847 \end{cases} \tag{8.2}$$

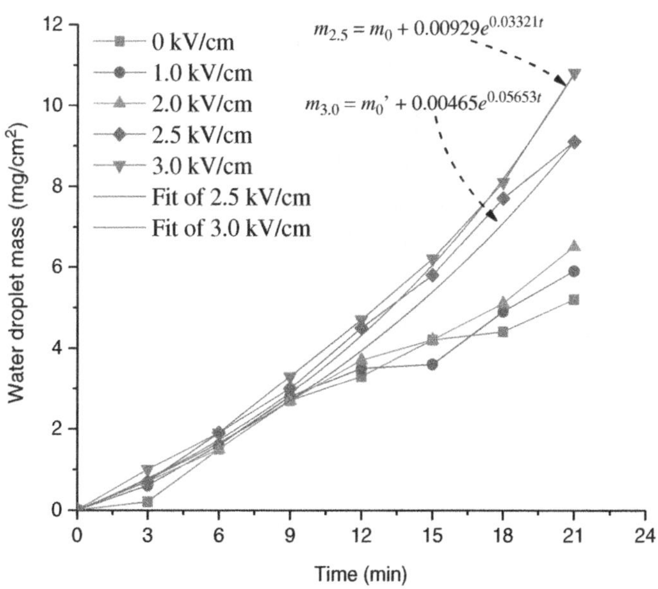

Figure 8.17 Relationship between water droplet mass and time with different electric fields ($T = 28\,°C$, $\Delta T = 13\,°C$, RH = 75%). *Source:* Wu et al. (2023)/John Wiley & Sons/CC BY 4.0.

The electric field distribution during the wetting process was simulated by COMSOL. The photos of water droplets were binarized first and then the mesh generation and calculation of the electric field are carried out, as shown in Figure 8.18.

In order to describe the degree of nonuniformity of the electric field, the distortion rate k was put into discussion. It was defined by the following expressions:

$$k = \frac{E_{\text{avg}}}{E_{\text{max}}} \tag{8.3}$$

$$E_{\text{avg}} = \frac{U}{\text{d}} \tag{8.4}$$

where E_{max} is the maximum electric field, E_{avg} is the average electric field, U is the voltage applied between electrodes, and d is the distance between two electrodes.

The evolution of the field distortion rate k during the wetting process is shown in Table 8.1. As the droplets increased in size, the electric field distortion became more pronounced.

In terms of the water droplet's proportion, Figure 8.19 demonstrated the change of k_A and k_L during the wetting process which can be summarized into three stages. Both parameters rose slowly at the beginning stage then increased rapidly and finally rose slowly again. The higher the electric field is, the larger the k_A and k_L are at the same wetting time with the same voltage applied.

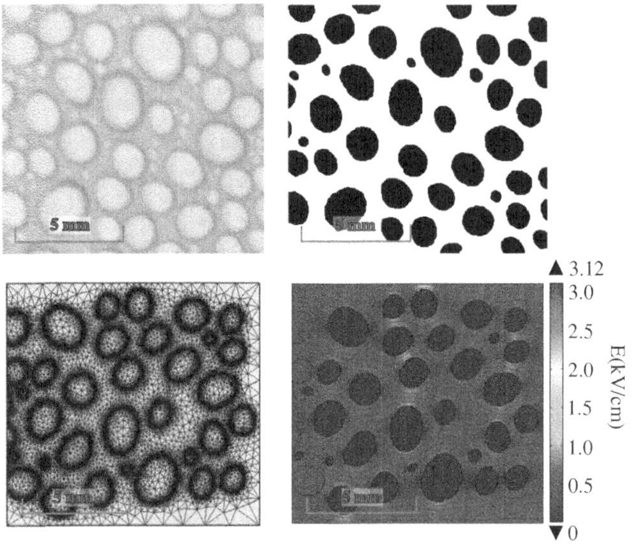

Figure 8.18 Method for the electric field calculation. *Source:* Wu et al. (2023)/John Wiley & Sons/CC BY 4.0.

Table 8.1 Field distortion rate k during the wetting process.

Wetting time (min)	Field distortion rate k
9	2.7444
15	3.1416
21	5.1644

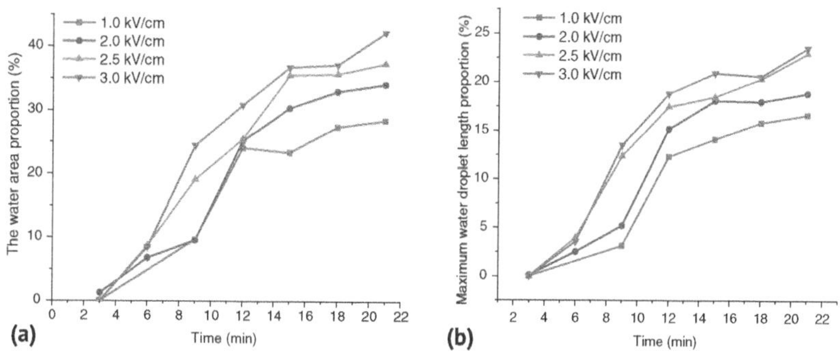

Figure 8.19 Change of water droplet proportion during the wetting process with different electric field: (a) area proportion of water droplets k_A; (b) lengthwise proportion along insulation path k_L. *Source*: Wu et al. (2023)/John Wiley & Sons/CC BY 4.0.

8.4.2 Flashover Characteristics

To study the wet flashover phenomena of the hydrophobic surface with temperature difference, the 80% dry flashover voltage of the test specimens (5.3 kV/cm) was applied at both ends of the sample. The same voltage supple and DAQ system in Figure 4.1 was used in this section. The test environment is 26 °C and RH 72%. The variable is the temperature difference which is set to 10 °C, 13 °C, and 16 °C. Once the hydrophobic surface's temperature was below the dew point, the small water droplets would gradually accumulate on the specimen. The flashover would finally occur but with a different withstand time and the time for the voltage applied to the flashover is recorded. The test result is shown in Figure 8.20. As the wetting speed increase (larger temperature difference), it will take less time for the occurrence of flashover.

Previous studies have shown that the flashover voltage of a hydrophobic surface covered by water is related to the proportion of water droplets and with higher water proportion, the flashover voltage is lower. The corresponding withstands times are circled in Figure 8.15, which pointed out that the flashover happened around the same value of k_L. It is found that it has a good correspondence which

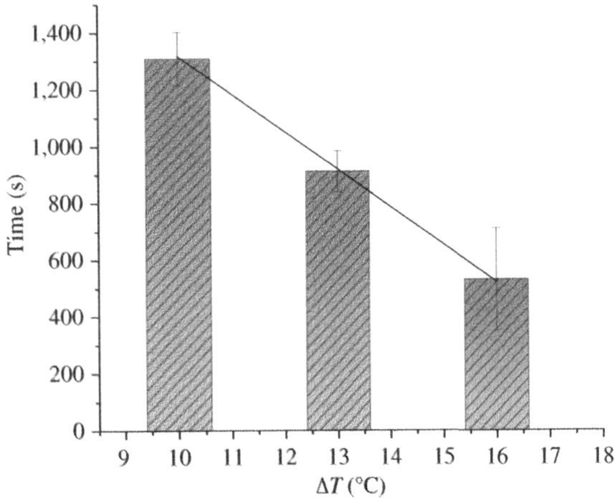

Figure 8.20 Relationship between temperature difference and withstand time.

proves that when the water droplet length proportion k_L reaches a threshold, the flashover will occur. the threshold is dominated by the residual insulation length. In withstanding test with 80% dry flashover voltage, the threshold of water proportion k_L for the occurrence of flashover should be around 20%. The experimental environment at that time (26 °C, RH 72%, 5.3 kV/cm) is not exactly the same as the parameters in Figure 8.13 (28 °C, RH 76%, 1.5 kV/cm). Therefore, the threshold is slightly bigger than 20% which is around 28%.

8.5 Summary

This chapter focused on the wetting behavior of hydrophobic surfaces under temperature gradients. By controlling the temperature difference and the electric field strength, the wetting characteristics (water mass and water morphology) of hydrophobic surface is studied by changing the temperature difference and electric field strength. The conclusions are as follows:

1. Water droplets begin to form on the hydrophobic surface only when its temperature falls below the dew point. Furthermore, a larger temperature difference results in a faster wetting rate.
2. The presence of an electric field enhances the wetting process on hydrophobic surfaces under temperature differences. Under an AC electric field, water droplets tend to coalesce, forming larger droplets that further distort the local electric field, thereby accelerating the wetting process. As a result, the curve representing water mass over time under high electric field strength exhibits an exponential trend.

3. When subjected to an 80% dry flashover voltage, the withstand time of the specimens decreases as the temperature difference increases. This is attributed to the fact that a larger temperature gradient accelerates droplet formation, reducing the time required for the droplet coverage ratio k_L to reach the flashover threshold.

Bibliography

Aouabed, F., Bayadi, A., and Boudissa, R. Flashover voltage of silicone insulating surface covered by water droplets under AC voltage, Electr. Power Syst. Res., 2017, 143, 66–72.

Aouabed, F., Bayadi, A., Rahmani, A.E. et al. Finite element modelling of electric field and voltage distribution on a silicone insulating surface covered with water droplets, IEEE Trans. Dielectr. Electr. Insul., 2018, 25 (2), 413–420.

Bojovschi, A., Rowe, W., and Wong, K. The influence of hanging water droplets on discharge activity, application to high voltage insulators, Appl. Phys. Lett., 2011, 98 (9), 3067.

Gouda, O.E. Flashover of overhead HVTL insulators under desert pollution conditions, Annual Conference on Electrical Insulation and Dielectric Phenomena, 1990, pp. 605–611.

Guan, Z.C., Wang, S.W., and Liang, X.D. Application and prospect of polymeric outdoor insulation in China, High Volt., 2000, 26 (6), 37–39.

Karady, G. The effect of fog parameters on the testing of artificially contaminated insulators in a fog chamber, IEEE Trans. Power Appar. Syst., 1975, 94 (2), 378–387.

Katada, K., Takada, Y., Takano, M. et al. Corona discharge characteristics of water droplets on hydrophobic polymer insulator surface, Proceedings of the 6th International Conference on Properties and Applications of Dielectric Materials (Cat. No.00CH36347), 2000, 2, pp. 781–784.

Li, Q., Rowland, S.M., Dupere, I. et al. The impact of water droplet vibration on corona inception on conductors under 50 Hz AC fields, IEEE Trans. Power Del., 2018, 33 (5), 2428–2436.

Liao, Y., Wang, Q., Yang, L., et al. Discharge behavior and morphological characteristics of suspended water-drop on shed edge during rain flashover of polluted large-diameter post insulator, Energies, 2021, 14 (6), 1652.

Liu, Y., Kong, X., Wu, Y. et al. Dynamic behavior of droplets and flashover characteristics for CFD and experimental analysis on SiR composites, IEEE Access, 2019, 7, 8095–8101.

Lopes, I.J.S., Jayaram, S.H., and Cherney, E.A. A study of partial discharges from water droplets on a silicone rubber insulating surface, IEEE Trans. Dielectr. Electr. Insul., 2001, 8 (2), 262–268.

Nazemi, M.H. and Hinrichsen, V. Experimental investigations on water droplet oscillation and partial discharge inception voltage on polymeric insulating surfaces under the influence of AC electric field stress, IEEE Trans. Dielectr. Electr. Insul., 2013, 20 (2), 443–453.

Que, W. and Sebo, S.A. Electric field and potential distributions along non-ceramic insulators with water droplets, Proceedings: Electrical Insulation Conference and Electrical Manufacturing and Coil Winding Conference (Cat. No.01CH37264), 2001, pp. 441–444.

Sarathi, R., Mishra, P., Gautam, R. et al. Understanding the influence of water droplet initiated discharges on damage caused to corona-aged silicone rubber, IEEE Trans. Dielectr. Electr. Insul., 2017, 24 (4), 2421–2431.

Song, S.Q., Wang, Y.Q., and He, Z.H. Experiment research on AC flashover performance of smearing indoor post insulator under condensation situation, Insul. Surge Arrest., 2016, (3), 25–29.

Wu, X., Cao, B., Li, Z. et al. DC breakdown characteristic of air gap with water droplets, IEEE Trans. Plasma Sci., 2021, 49 (6), 1962–1968.

Wu, X., Cao, B., Wang, L. et al. Wetting characteristics of the hydrophobic material surface through condensation, High Volt., 2023, 8 (1), 31–37.

Wu, X., Yin, F., Cao, B. et al. Research on flashover characteristics of dry band-water band on hydrophobic surface, CSEE J. Power Energy Syst., 2021, 10 (1), 392–400.

Xiao, M., Wen, C., and Sima, W.X. Discharges between discrete water drops on the silicone rubber surface and reason unknown flashover, J. Chongqing Univ., 2006, 29 (6), 42–46.

Xu, J.Y. 110~500 kV analysis and preventive measures for synthetic insulator flashover on 110~500kV lines, Electr. Power, 2000, 33 (5), 54–56.

Yang, F.F. Experimental study on wet flashover and condensation prevention of biomimetic super hydrophobic materials. M.D. thesis, Huazhong University of Science and Technology, 2020.

Zhang, C., Wang, L., and Guan, Z. Investigation of DC discharge behavior of polluted porcelain post insulator in artificial rain, IEEE Trans. Dielectr. Electr. Insul., 2016, 23 (1), 331–338.

9

Flashover Characteristics of Hydrophobic Surfaces with Different Wetting Degree

9.1 Introduction

As previously discussed, surface properties—particularly wettability—play a decisive role in pollution flashover behavior under various environmental and electrical stresses. Hydrophobicity, the ability of a surface to repel water, has emerged as a key factor influencing surface discharge characteristics. Hydrophobic surfaces hinder the formation of continuous water films, which would otherwise facilitate leakage currents and promote surface discharges.

However, hydrophobicity is not a binary attribute. Instead, it covers the entire transition process from completely hydrophilic to completely hydrophobic, as defined by the wetting degree. Understanding how different degrees of surface hydrophobicity influence flashover performance is critical for the design and selection of insulation materials in high-voltage applications, particularly under polluted or moist conditions.

This chapter investigates the flashover characteristics of hydrophobic surfaces across a range of wetting degrees. It aims to provide insight into how surface wettability impacts the initiation and progression of flashover events. The findings not only deepen our understanding of flashover mechanisms but also support the development of practical strategies—such as surface modification and material selection—for enhancing the reliability of high-voltage insulation systems.

For the surface flashover on the hydrophobic surface with water, this chapter has designed a series of experiments to find out the experimental phenomena and mechanisms on the hydrophobic surface. The experiments were divided into two parts. First, with a low proportion of water, the surface flashover with water droplets was studied by changing the water droplets' parameters (arrangement, volume, conductivity). Secondly, the flashover phenomenon with a high proportion of water was studied as well. In this case, the water bands with different lengths, widths, and conductivity were placed to find out their influence on surface flashover. On this basis, the arc generation and development process of the surface covered with "water droplet and water band" is analyzed.

9.2 Flashover Characteristics with Lower Wetting Degree

9.2.1 Experimental Arrangement

In the first case, only a limited amount of water is present on the hydrophobic surface, resulting in the formation of discrete water droplets rather than a continuous film. The rest of the surface remains relatively dry and clean due to its water-repellent nature. As a result, the surface condition can be described as a composite interface consisting of individual water droplets interspersed with air and dry surface regions.

To investigate the influence of these water droplets on flashover behavior, systematic variations were introduced in the following experiments. Specifically, the size, number, and conductivity of the droplets were adjusted to assess their individual and combined effects on surface discharge initiation and development.

The experimental setup is illustrated in Figure 9.1. A 50 kV/100 kVA transformer was employed to generate the high voltage required for the tests. The applied

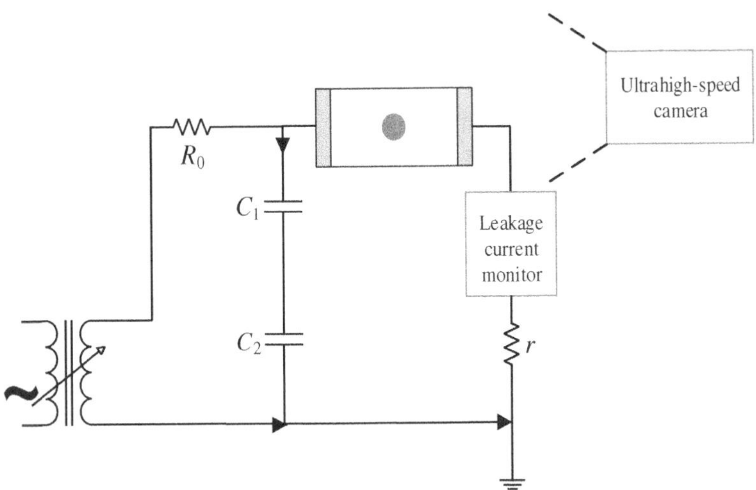

Figure 9.1 Experiment platform.

voltage between the two electrodes was measured using a capacitive voltage divider, ensuring accurate monitoring of the test conditions. To capture electrical activity on the surface, the leakage current was recorded using a data acquisition (DAQ) board operating at a sampling rate of 5 kSamples/s. In order to observe the dynamic flashover process in detail, a Phantom v2012 high-speed camera was positioned in front of the specimen. This camera is capable of recording up to 22,500 frames per second (fps) at a resolution of 1280 × 800 pixels, allowing for high-resolution visualization of rapid discharge events.

Silicone rubber tends to undergo physical and chemical degradation under prolonged surface discharge, leading to a gradual loss of hydrophobicity. To better observe the physical processes during flashover under controlled laboratory conditions—and to maintain good hydrophobicity across multiple test cycles—polytetrafluoroethylene (PTFE) was selected as the test material in this section. PTFE exhibits flashover characteristics similar to those of silicone rubber, making it a suitable substitute for repeatable and reliable testing.

For better visualization of the water droplets, indigo blue dye was added to enhance contrast. The conductivity of the water was adjusted using sodium chloride (NaCl), with conductivity levels ranging from 20.1 μS/cm to 110.2 mS/cm to simulate different pollution conditions. The test sample was a PTFE board with dimensions of 7.0 cm × 5.0 cm, and copper foil electrodes were affixed on two opposite sides, providing an insulation length of 5.0 cm between them. Water droplets were arranged in three configurations—a single droplet, a pair of droplets, and four droplets—as illustrated in Figure 9.2.

9.2.2 Effect of Water Conductivity and Volume

A single water droplet was placed at the center of the PTFE sample's surface. The voltage was gradually increased at a rate of 1.4 kV/s until flashover occurred. The entire flashover process, including changes in the droplet shape and the arc development, was captured and is presented in Figure 9.3.

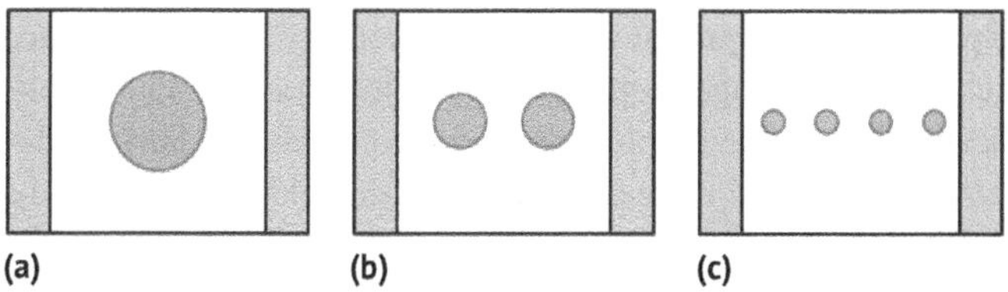

Figure 9.2 Water droplets arrangements: (a) single droplet; (b) double droplets; (c) four droplets.

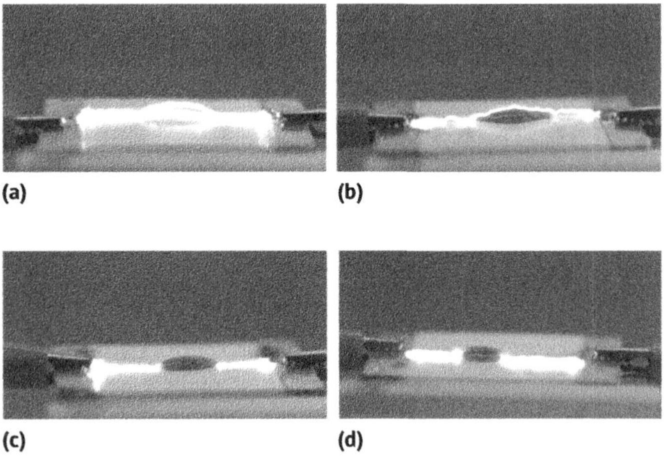

Figure 9.3 Arc patterns of water droplets with different conductivity: (a) 20.1 μS/cm; (b) 426.5 μS/cm; (c) 17.5 mS/cm; (d) 110.2 mS/cm.

It can be observed that the arc position during flashover on a hydrophobic surface containing water droplets is strongly influenced by the conductivity of the water. When the water conductivity is low, the arc tends to bridge the droplet from above, creating a bypass that avoids direct current flow through the droplet. In contrast, at higher water conductivity, the arc interacts more directly with the droplet, forming a series connection, where the arc current passes through the droplet itself.

In other words, the conductivity of the water droplet determines the electrical path of the arc—either as a parallel connection (arc bridging over the droplet) in the case of low conductivity, or a series connection (arc passing through the droplet) when the conductivity is relatively high. This behavior highlights the critical role of droplet conductivity in shaping the flashover development mechanism on hydrophobic surfaces.

The test results for single and double water droplets attached to the hydrophobic surface are presented in Figures 9.4 and 9.5, respectively. As demonstrated in both figures, the flashover voltage decreases with an increase in water droplet volume. This indicates that larger droplets are more conducive to surface discharge development, likely due to increased local electric field distortion and a larger conductive path.

However, it is noteworthy that the flashover voltage does not show a consistent correlation with the conductivity of the water droplets. This suggests that, within the tested range, droplet volume plays a more dominant role than conductivity in influencing the onset of flashover on hydrophobic surfaces.

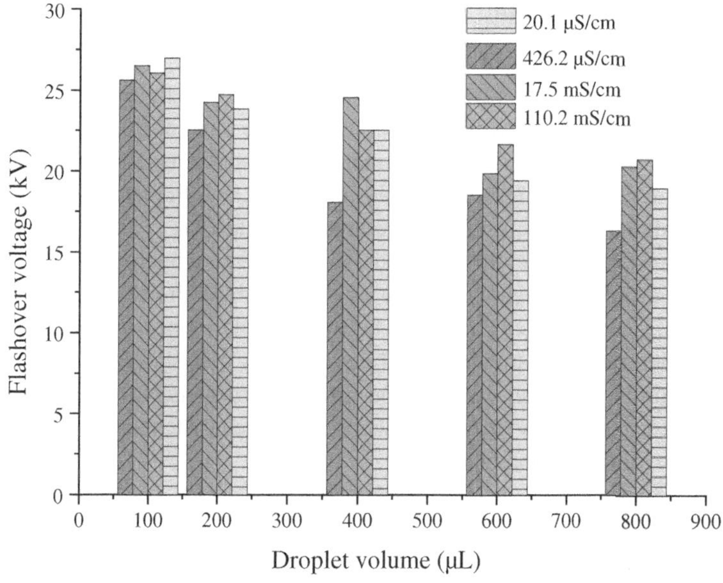

Figure 9.4 Relationship between flashover voltage and single droplet's volume.

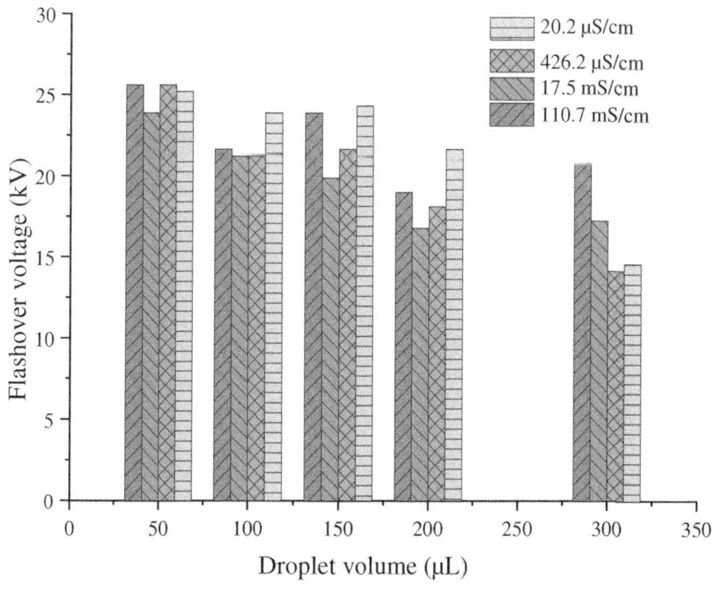

Figure 9.5 Relationship between flashover voltage and double droplets' volume.

9.2.3 Effect of Droplet Length

To account for the deformation of the water droplet under the influence of the electric field, image processing software was employed to analyze the droplet shape immediately before flashover. Specifically, the length of the water droplet just prior to flashover was measured and used as a parameter to investigate its effect on flashover behavior. The relationship between flashover voltage and droplet length was subsequently analyzed, and the results are presented in Figure 9.6.

Experimental results indicate that the flashover voltage on a hydrophobic surface with water droplets decreases linearly as the length of the water droplet increases. This suggests a direct relationship between droplet elongation—caused by electric field deformation—and increased susceptibility to flashover. However, with a greater number of water droplets, the variability (dispersibility) in flashover voltage measurements also increases, likely due to inconsistencies in droplet placement, shape, or interaction effects.

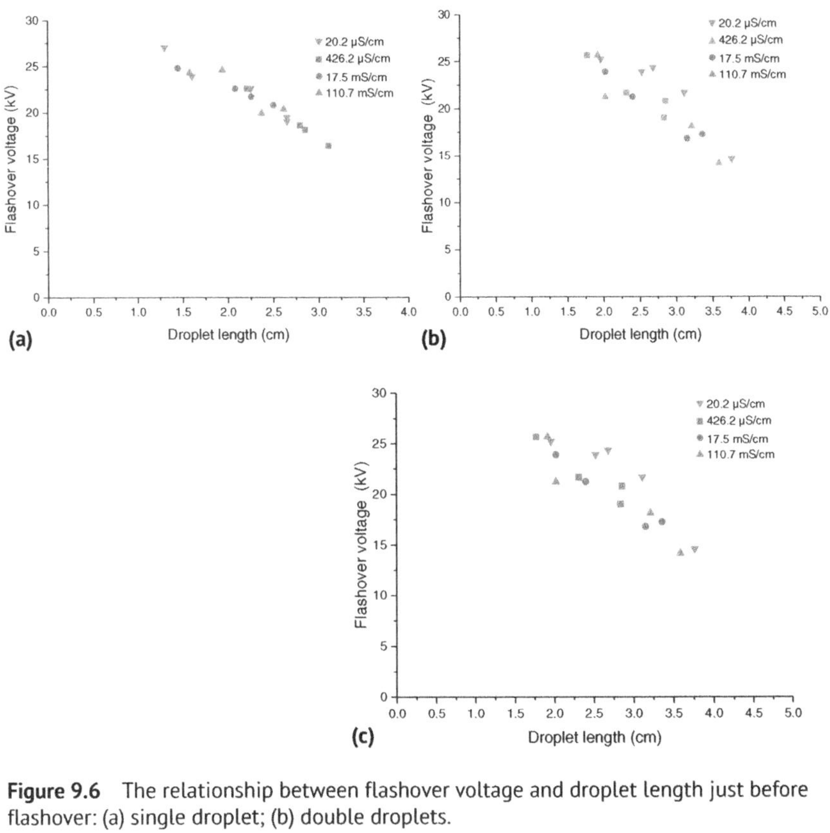

Figure 9.6 The relationship between flashover voltage and droplet length just before flashover: (a) single droplet; (b) double droplets.

To further investigate the influence of water conductivity on flashover voltage—while eliminating variables introduced by liquid deformation—copper foil pieces of different sizes were used to simulate droplets. These copper elements were arranged in a configuration closely matching the water droplet patterns shown in Figure 9.7, allowing for a controlled comparison between conductive solid contacts and actual water droplets.

The arc development path when using copper foil is illustrated in Figure 9.8. Due to copper's excellent electrical conductivity, the arc tends to connect directly through the copper foil rather than bridging over it, forming a clear series connection. This behavior contrasts with that observed in some water droplet cases, where bridging may occur depending on the conductivity.

As shown in Figure 9.9, the flashover voltage exhibits a linear dependence on the length of the copper foil, consistent with the trend observed for water droplet length in Figure 9.6. This similarity reinforces the conclusion that the length of the conductive

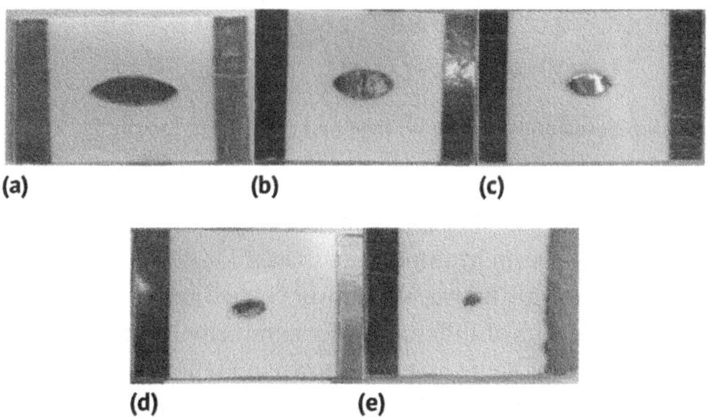

Figure 9.7 The photos of copper foil with different lengths: (a) 4 cm copper; (b) 3 cm copper; (c) 2 cm copper; (d) 1 cm copper; (e) 0.5 cm copper.

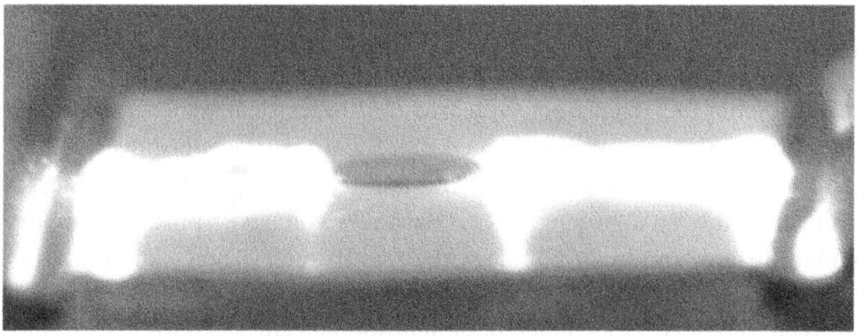

Figure 9.8 The flashover phenomenon with copper foil.

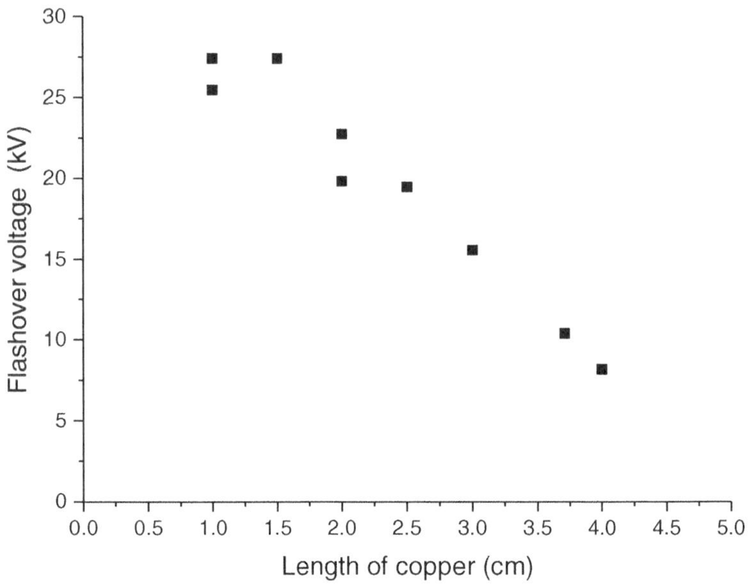

Figure 9.9 The relationship between flashover voltage and copper foil length.

path—regardless of whether it is a liquid or solid conductor—plays a dominant role in determining the flashover voltage on hydrophobic surfaces.

In order to find out the relationship between flashover voltage and water droplet length, the flashover voltage results of different water conductivity and different number of droplets is plotted in the same graph as shown in Figure 9.10. It can be seen that the flashover voltage with water droplets is depended on the length of droplets. Meanwhile, the water conductivity has no clear influence on flashover voltage. The length of the dry area along the x axis is defined as dry band length l_{dry} and the droplets length along x axis just before flashover is defined as l_{water}. The length of the dry area can be represented as:

$$l_{dry} = L - l_{water} \tag{9.1}$$

where L represents the length between two electrodes (leakage length) which is 5.0 cm. The previous research indicated that the dry surface flashover voltage is smaller than that of the dry air case. And the dry surface flashover voltage is directly linear to the leakage length. The flashover voltage of pure air gap, dry hydrophobic surface, and hydrophobic surface with water droplets are placed in the same graph as Figure 9.11.

As observed, the flashover voltage on a hydrophobic surface with water droplets is close to that of a completely dry surface, and both are slightly lower than

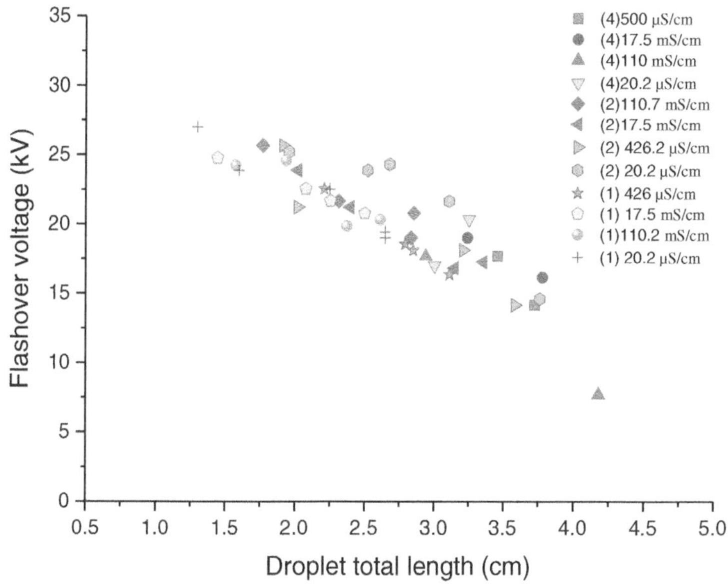

Figure 9.10 The relationship between flashover voltage and droplets total length.

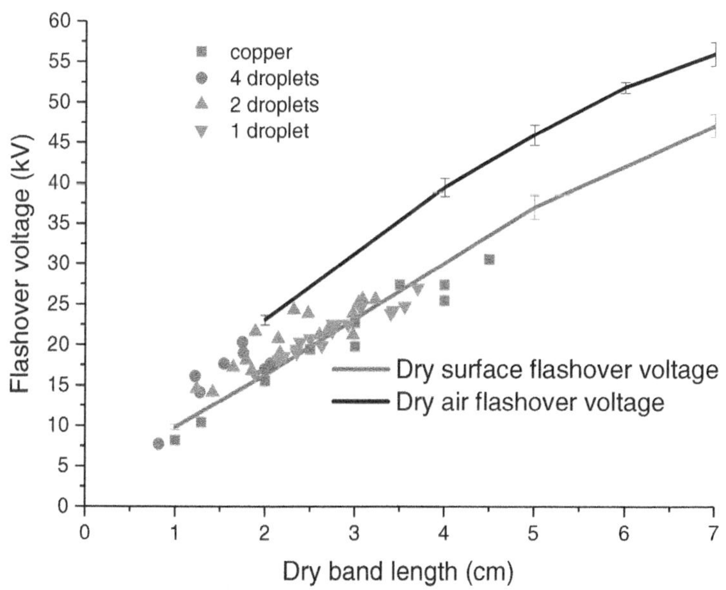

Figure 9.11 The flashover voltage of air, dry hydrophobic surface and hydrophobic surface with water droplets. *Source*: Cao et al. (2021)/with permission of IOP Publishing Ltd.

the flashover voltage across a pure dry air gap. This suggests that the presence of isolated water droplets on a hydrophobic surface does not significantly reduce the flashover voltage, unless a conductive path—such as a broken dry band—is established. Once this dry band breaks down, flashover is likely to occur.

However, results from artificial pollution flashover tests show that the degree of contamination still affects flashover voltage, even when the surface exhibits strong hydrophobicity. This appears to contradict the findings from the water conductivity experiments presented earlier. The key difference lies in the test conditions: in artificial pollution tests on composite insulators, a portion of the surface-deposited salt dissolves into the moisture layer, forming a more conductive path, despite the material's hydrophobic properties.

More importantly, the proportion of the surface covered by water is substantially different between the two scenarios. In this section's experiments, the water coverage on the insulation length is relatively small, limited to discrete droplets. In contrast, artificial pollution tests often result in extensive water coverage, forming a near-continuous conductive film. This fundamental difference in water distribution and coverage is likely the reason for the differing flashover behavior.

In the next section, the focus will shift to analyzing flashover performance under conditions with a higher water coverage ratio, to further investigate the impact of water distribution on hydrophobic surfaces.

9.3 Flashover Characteristics with Higher Wetting Degree

Due to the differences in surface properties between composite insulators and inorganic insulators, the physical processes of surface flashover on these two types of materials also differ significantly. The flashover mechanism on composite insulators, or more generally on hydrophobic surfaces, can be summarized as follows:

1. Under the influence of environmental factors such as light rain, fog, or condensation, the insulator surface becomes covered with discrete water droplets.
2. When subjected to an electric field, these droplets may distort, move, split, or merge, potentially leading to the formation of intermittent water bands.
3. A composite discharge path consisting of both dry bands and water bands is formed, and the arc tends to develop along or near the water bands, progressing rapidly across the surface.

This process is visually illustrated in Figure 9.12, which shows the actual flashover development on a hydrophobic surface. To investigate this phenomenon under controlled laboratory conditions—and to preserve stable hydrophobicity over multiple tests—a PTFE board was used as the test sample in this section, just as

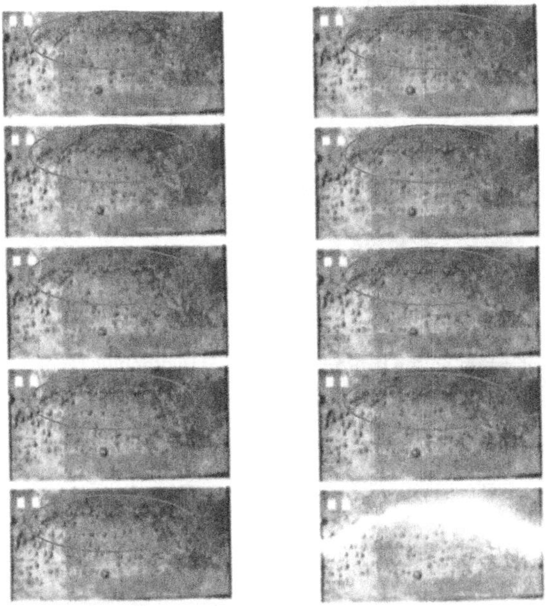

Figure 9.12 Actual flashover process on hydrophobic surface.

in the previous experiments. PTFE was chosen because it exhibits flashover characteristics similar to those of silicone rubber, making it a suitable substitute for high-temperature vulcanized silicone rubber insulators during repeated testing.

9.3.1 Experimental Arrangement

The PTFE specimen used in this experiment had dimensions of 35.0 cm × 5.0 cm. Due to the material's excellent hydrophobicity, it was not feasible to directly apply a continuous water band of the desired shape and size onto the surface. To overcome this, a hydrophilic tape with a predefined shape was affixed to the PTFE surface, enabling controlled formation of a water band. The length of the hydrophilic tape could be varied to suit different test configurations.

For the electrode setup, two copper strips measuring 5.0 cm × 1.0 cm × 0.2 cm were used. One strip was fixed at the edge of the specimen surface, while the other was mounted on a movable platform, allowing adjustment of the electrode spacing as needed. The complete arrangement of the specimen and electrodes is illustrated in Figure 9.13.

Prior to testing, water with the desired conductivity was prepared by dissolving sodium chloride in deionized water. The resulting solution had a conductivity range from 100 to 500 μS/cm. This water was then applied to the hydrophilic tape

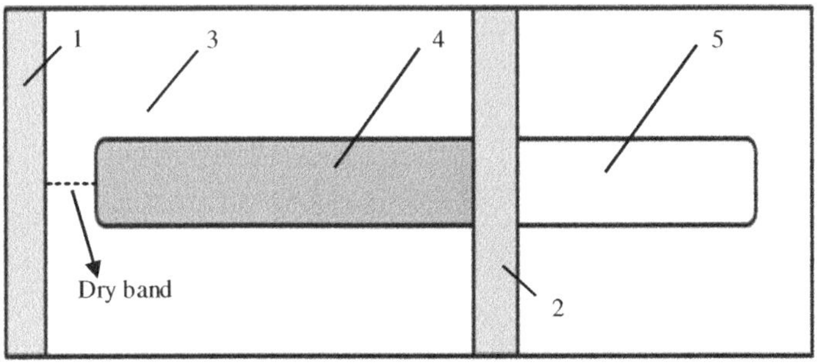

Figure 9.13 Specimen arrangement. (1: Fixed electrode; 2: Sliding electrode; 3: PTFE specimen; 4: Water band (water film on the hydrophilic tape); 5: Hydrophilic tape).

Table 9.1 Test results of different water band lengths with conductivity of 100 μS/cm.

Length (cm)	Test result
4.0	Flashover
4.5	Flashover
5.0	Flashover
5.5	Flashover
6.0	Withstand
6.5	Withstand
7.0	Withstand

positioned between the two copper electrodes, forming a controlled water band. To enhance visibility during observation, the water was dyed with indigo.

In each test, the voltage between the electrodes was increased at a rate of 1.0 kV/s until it reached 1.05 times the dry band flashover voltage. For the specific tests discussed here, the applied voltage was fixed at 13.2 kV, corresponding to the 1.05× multiplier. The dry band length was kept constant at 1.0 cm, while other parameters were standardized: the water band thickness was maintained at 1.0 cm, the height at 0.1 cm, and the water conductivity at 100 μS/cm.

By adjusting the position of the movable copper electrode, the length of the water band was varied from 4.0 to 7.0 cm to evaluate its influence on flashover performance. The detailed experimental results are summarized in Table 9.1.

When the length of the water band listed in Table 9.1 was less than 6.0 cm, breakdown of the dry band consistently led to a complete flashover of the specimen surface. However, when the water band length reached 6.0 cm or greater, although

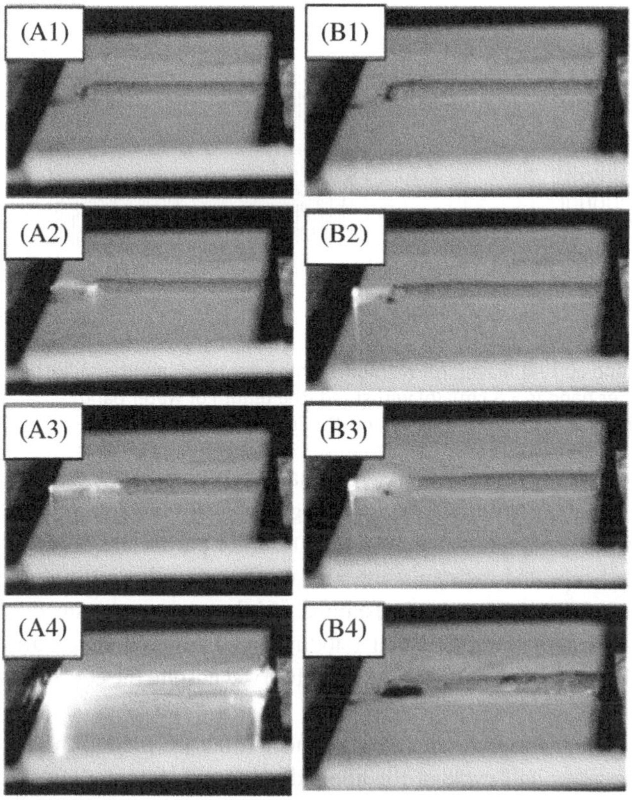

Figure 9.14 The flashover and extinction phenomenon on hydrophobic surface. *Source:* Cao et al. (2021)/with permission of IOP Publishing Ltd. Note: The length of water band in A1–A4 is 5.5 cm and the time interval of A1–A4 is 100 ms; the length of water band in B1–B4 is 6.0 cm and the time interval is 400 ms.

the dry band still broke down, only a partial arc was observed. This partial arc repeatedly extinguished and reignited without propagating further along the surface. Over time, the arc-induced heating caused evaporation of the water band, preventing the occurrence of a full flashover.

Figure 9.14 illustrates the arc development process captured by a high-speed camera for water band lengths of 5.5 and 6.0 cm, respectively. For the 5.5 cm water band, flashover occurred approximately 200 to 300 ms after the initial appearance of the partial arc. In contrast, the 6.0 cm water band experiment showed the partial arc failing to develop into a full flashover.

Simultaneously, current measurements were recorded during both experiments. Figure 9.15 displays the measured current waveforms for the two water band lengths. In Figure 9.15a, corresponding to the 5.5 cm water band, the partial

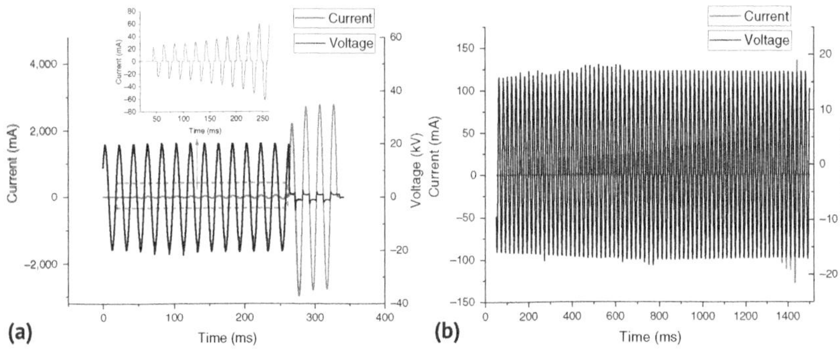

Figure 9.15 Voltage and current waveform with different water band length: (a) flashover (the length of water band is 5.5 cm); (b) withstand (the length of water band is 6.0 cm). *Source*: Cao et al. (2021)/with permission of IOP Publishing Ltd.

arc initiated at around 45 ms, with the current gradually increasing until a sudden spike to approximately 3,000 mA at 263 ms, marking the flashover event.

For the 6.0 cm water band, shown in Figure 9.15b, the partial arc began at 112 ms and exhibited repeated extinguishing and reignition cycles until about 585 ms. After this point, the arc remained sustained and the current rose steadily, reaching a peak of approximately 135 mA at 1,400 ms, after which the arc extinguished permanently. No further partial arcs were observed despite the continued application of voltage.

It should be noted that the extinguishing of the partial arc predominantly occurs during the negative half-cycle of the applied voltage in the time interval from 112 to 585 ms. This observation aligns with the established fact that the breakdown voltage of the dry band is higher during the negative cycle compared to the positive cycle, making arc maintenance more difficult in the negative half-cycle.

To quantify the threshold beyond which flashover becomes highly likely along the specimen surface for a given dry band length, a Critical Flashover Length of the Water Band (CFLWB) is defined. The CFLWB is calculated as the average of two adjacent water band lengths where experimental results differ—specifically, where one length leads to flashover and the next does not, or vice versa. Based on this definition, the CFLWB reported in Table 9.1 is 5.75 cm.

9.3.2 Effect of Water Conductivity

Using the same method described previously, experiments were conducted to investigate the effect of water conductivity on the CFLWB. During these tests, the width and height of the water band and the length of the dry band were kept

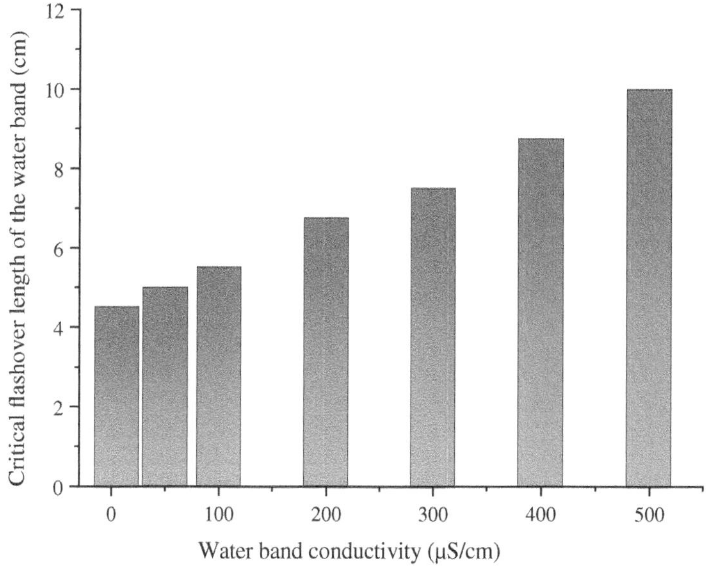

Figure 9.16 Relationship between the CFLWB and water conductivity.

constant at 1.0 cm, 0.1 cm, and 1.0 cm, respectively. The length of the water band was varied to determine the CFLWB at different water conductivities.

The results, shown in Figure 9.16, indicate a strong linear relationship between CFLWB and water conductivity. Notably, when the water band consisted of deionized water with a conductivity of 5 μS/cm, the CFLWB was found to be 4.5 cm.

9.3.3 Effect of Dry Band Length

To investigate the effect of dry band length, the water conductivity, width, and height of the water band were kept constant at 100 μS/cm, 1.0 cm, and 0.1 cm, respectively. The experiments tested dry band lengths ranging from 1.0 to 3.0 cm, with the applied voltage between the copper electrodes set at 1.05 times the dry band flashover voltage. Figure 9.17 shows the variation of the CFLWB with different dry band lengths. As illustrated, the CFLWB initially increases linearly as the dry band length increases, but after reaching a certain point, the rate of increase slows down slightly. From Figure 9.18, it can be concluded that the flashover voltage increases as the insulation distance becomes longer. Additionally, the flashover voltage across the dry band is slightly lower than that of a pure air gap with the same distance.

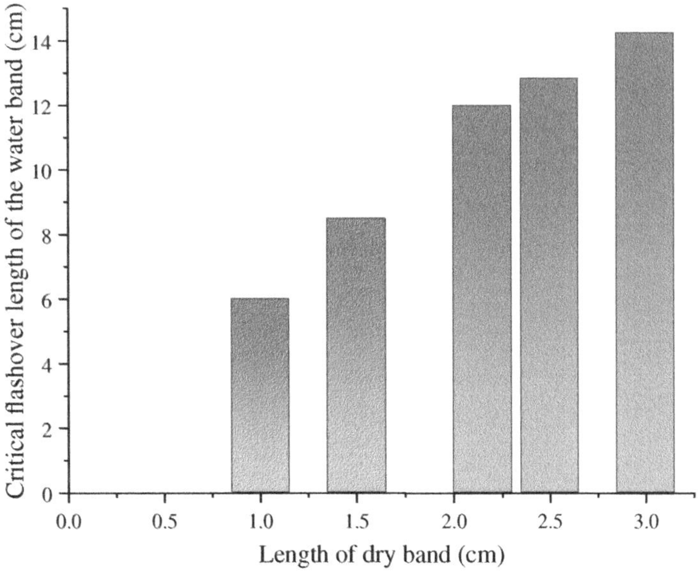

Figure 9.17 Relationship between the CFLWB and the length of the dry band.

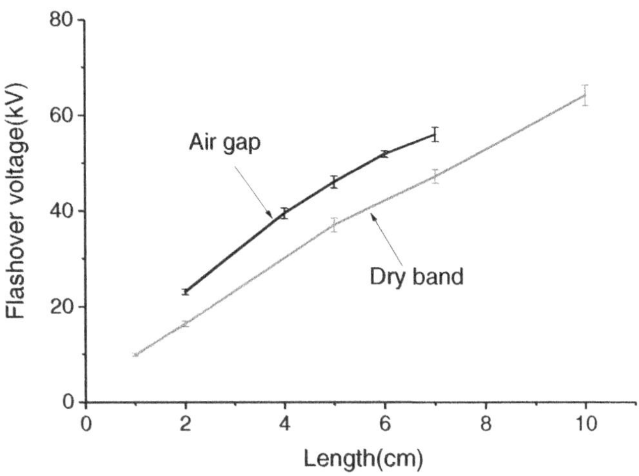

Figure 9.18 The relationship between the flashover voltage and the length of the dry band.

9.3.4 Effect of Water Band Width

To study the effect of water band width on the flashover characteristics, the water conductivity, height of the water band, and length of the dry band were maintained at 100 μS/cm, 0.1 cm, and 1.0 cm, respectively. The CFLWB values obtained for

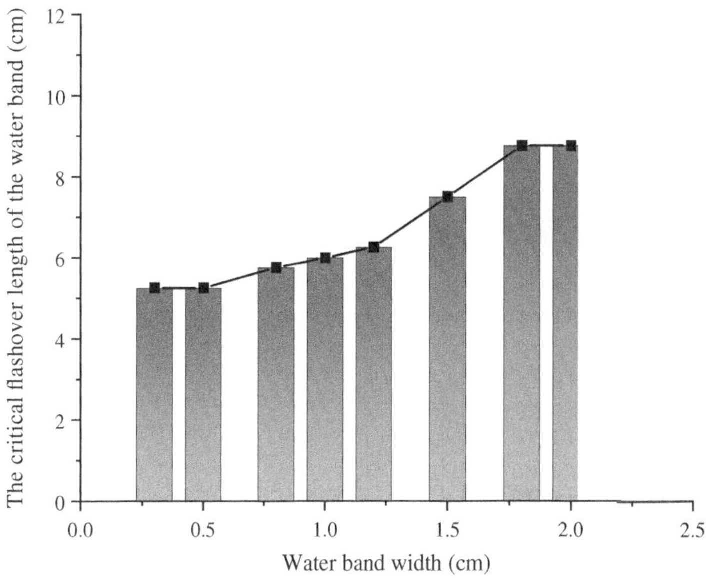

Figure 9.19 The relationship between the CFLWB and the water band width.

varying water band widths, ranging from 0.25 to 2.0 cm, are presented in Figure 9.19. As the width of the water band increases, the CFLWB initially rises but then levels off, indicating a saturation effect.

9.4 Discussion of Surface Flashover with Water Involved

Due to the AC voltage applied between the electrodes, the arc polarity alternates in each cycle. An illustrative example of the arc during both the positive and negative cycles is shown in Figure 9.20. It can be clearly observed that during the positive half-cycle, the arc exhibits several distinct bright spots, while during the negative half-cycle, it manifests as a uniform glow discharge. This phenomenon aligns closely with observations previously reported by researchers studying the characteristics of arc roots on electrolyte surfaces.

Considering that the water band is connected between the electrodes, the test specimen essentially forms an "electrode–dry band–water band–electrode" circuit. Therefore, the overall circuit resistance can be derived directly from the measured current-voltage waveform. With the lengths of the water band and dry band fixed at 10.0 and 1.0 cm, respectively, it was found that although the polarity alternation significantly affects the arc shape, it has a minimal influence on the overall circuit resistance, as depicted in Figure 9.21.

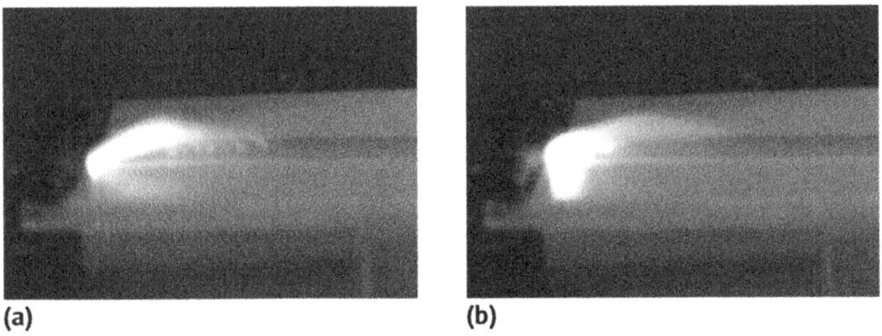

Figure 9.20 The arc shape in the positive and the negative cycle: (a) positive arc; (b) negative arc.

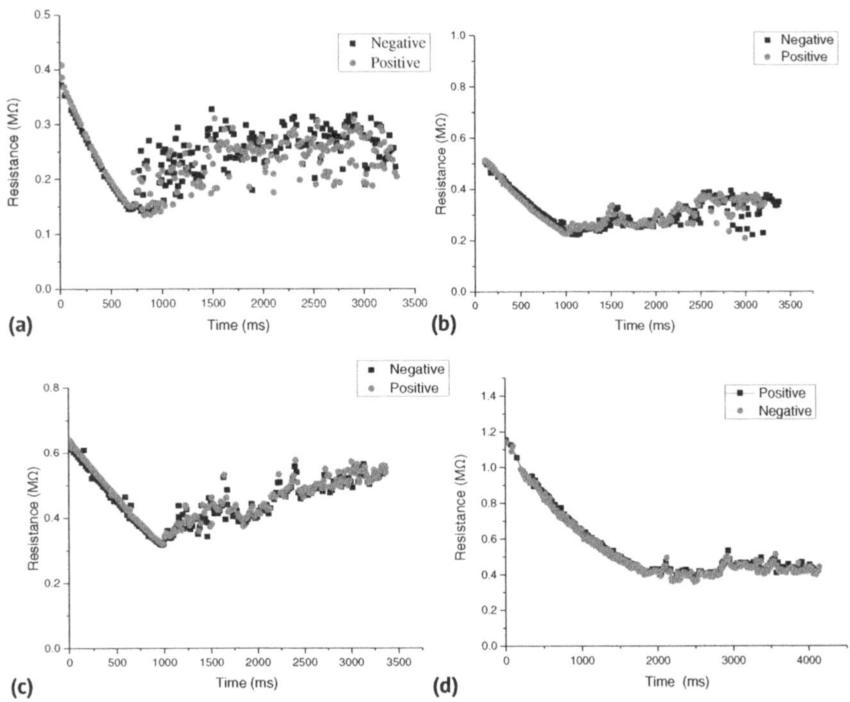

Figure 9.21 Influence of arc current's polarity (water band length = 10.0 cm and dry band length = 1.0 cm): (a) 400 μS/cm; (b) 300 μS/cm; (c) 200 μS/cm; (d) 100 μS/cm.

Moreover, it is clear that the observed arc root does not directly correspond to the actual current density distribution at the water band surface near the arc root, indicating a need for further investigation into this relationship. Since the heat generated by the leakage current will gradually evaporate the water band, it is important to analyze both the heat produced during discharge and its subsequent influence on the circuit's resistance. In this experiment, the dry band length was fixed at 1.0 cm, and the dimensions of the water band were 9.0 cm × 1.0 cm × 0.1 cm. When the applied voltage was increased to 1.05 times the breakdown voltage, a partial arc formed but was unable to propagate further, as illustrated in Figure 9.22. The instant at which the first arc current appeared is designated as time zero. Additionally, the leakage current was recorded using a DAQ system, and the resulting data are presented in Figure 9.23.

Figure 9.22 Water band and arc during discharge.

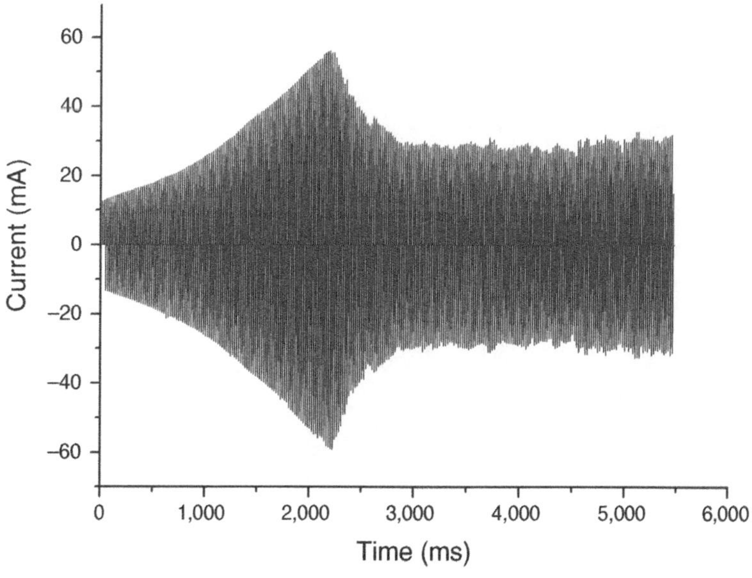

Figure 9.23 Current waveform during discharge.

Clearly, as discharge time increases, the amplitude of the arc current continues to rise despite the arc being unable to propagate forward. After approximately 2,228 ms of continuous discharge, the amplitude of the current begins to decrease, stabilizing at a lower value until the power is switched off. As illustrated in Figure 9.22, after 2,401.2 ms from the initial arc occurrence, significant heat generation even leads to boiling of the water band. The resulting air bubbles within the water band further increase its resistance.

9.4.1 Arc Generation on the Surface

To analyze the electric field distribution near the dry band–water band interface, simulations were conducted. The simulation model was built to replicate the physical test setup accurately, with results presented in Figure 9.24. It is clear from the figure that the electric field distortion is concentrated at the boundary between the dry band and the water band. This occurs because the relative permittivity of air ($\varepsilon_{air} \approx 1$) is significantly lower than that of water ($\varepsilon_{water} \approx 80$), causing the applied voltage to predominantly concentrate across the smaller capacitance of the dry band.

In this section, the applied test voltage was consistently set to 1.05 times the dry band flashover voltage, ensuring the initiation of an arc in each test. However, whether a complete flashover occurs depends on the subsequent propagation and development of the partial arc following the initial breakdown of the dry band.

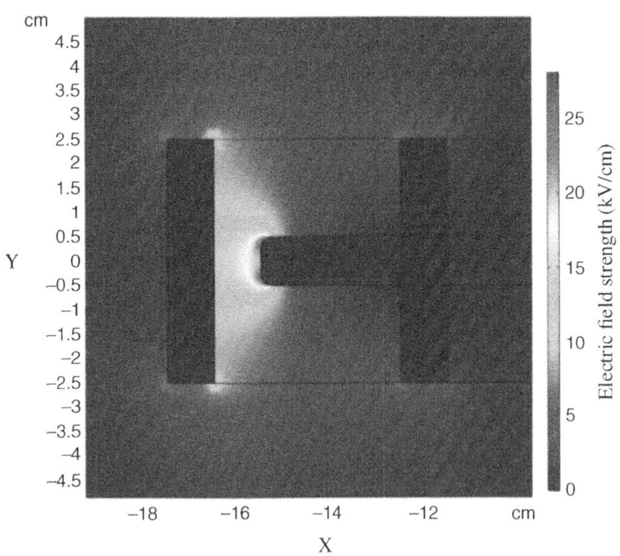

Figure 9.24 Electric field distribution around dry band-water band.

Figure 9.25 The schematic of water band arc.

Figure 9.25 represents the flashover arc model, the circuit is in a series connection with arc and residual resistance.

As it can be seen that the dry band-water band flashover model is similar to the Obenaus model:

$$U = \mathrm{A}xI^{-\mathrm{n}} + R(x)I \tag{9.2}$$

where U, I represent the voltage and leakage current, x represents the length of the arc, and $R(x)$ represents the remaining resistance of the pollution layer. A and n are the static arc characteristic constant.

The critical voltage is deduced based on the above flashover model:

$$U_c = \left(1 + \frac{1}{n}\right)(nx_c\mathrm{A})^{\frac{n}{n+1}} R(x_c)^{\frac{n}{n+1}} \tag{9.3}$$

where U_c represents the critical flashover voltage, X_c represents the critical length of the arc. Dr. Wilkins has given the formula for remaining resistance of pollution layer:

$$R(x) = \frac{1}{2\pi\gamma}\left[\frac{2\pi(L-x)}{a} + \ln\frac{a}{2\pi r_{\mathrm{arc}}}\right] \tag{9.4}$$

where L represents the leakage length, γ represents the pollution layer's conductivity, a is the width of the pollution layer and r_{arc} represents the radius of the arc.

Substituting Eq. (9.4) into Eq. (9.3) yields:

$$U_c = \left(1 + \frac{1}{\mathrm{n}}\right)(\mathrm{nA}x)^{\frac{1}{\mathrm{n}+1}}\left\{\frac{1}{2\pi\gamma}\left[\frac{2\pi(L-x)}{a} + \ln\frac{a}{2\pi r_{\mathrm{arc}}}\right]\right\}^{\frac{\mathrm{n}}{\mathrm{n}+1}} \tag{9.5}$$

Only considering the effect of γ, and other parameters are given, Eq. (9.5) can be rewritten as a formula of L and γ.

$$L = a \cdot \left(\frac{Uc}{\left(1 + \frac{1}{\mathrm{n}}\right)(\mathrm{nA}x)^{\left(\frac{1}{\mathrm{n}}\right)}}\right)^{\left(\frac{(\mathrm{n}+1)}{\mathrm{n}}\right)} \cdot \gamma + x - \frac{a}{\pi}\ln\frac{a}{2\pi r_{\mathrm{arc}}} \tag{9.6}$$

It can be seen from Eq. (9.6) that L and γ are linear dependent which is consistent with the result of Figure 9.16. U_c is the necessary condition but not sufficient for flashover. In order to discuss the arc's development mechanism, the prerequisites of the flashover on the hydrophobic surface are conducted in the following section.

9.4.2 Arc Development Criterion

The experimental results presented above demonstrate that under AC voltage conditions, flashover on hydrophobic surfaces is strongly influenced by the development of the partial arc. Given a specific dry band, flashover may not necessarily occur even if partial arcs have formed, particularly when the water band is sufficiently long or the water conductivity is relatively low. Thus, to fully understand the flashover mechanism on hydrophobic surfaces, it is critical to study the detailed process of arc development.

Previous studies have explored the propagation mechanisms of arc roots across naturally contaminated insulator surfaces using polluted strips as simulation tools. Shkuropat concluded that arc propagation could continue forward as long as the current increased. Hampton proposed a criterion based on voltage gradients, suggesting that arc propagation occurs when the arc voltage gradient is lower than that of the residual surface resistance. R. Wilkins emphasized that arc development fundamentally involves ionization and the regeneration of the arc root. While these theories explain specific phenomena, a comprehensive understanding of arc development on polluted layers remains incomplete.

According to the experimental results presented, significant differences between water band–dry band discharges on hydrophobic and hydrophilic surfaces can be summarized as follows:

1. On a hydrophobic surface, the arc closely follows the water band along the surface, whereas on a hydrophilic surface, thermal buoyancy often causes the arc to float above the dry area during propagation.
2. Arc development is faster on the water band of a hydrophobic surface compared to a hydrophilic one. When surface hydrophobicity decreases, the resultant water band is notably thicker, preventing the formation of a dry zone at the arc root.
3. During partial arc initiation and propagation, the liquid layer on a hydrophilic surface rapidly evaporates due to heat generated by the current, causing increased resistance near the arc root. Conversely, on a hydrophobic surface, the water band initially experiences a decrease in resistance followed by a gradual increase, attributable to its greater thickness.

Once a partial arc forms, differences in surface resistance significantly influence the conditions necessary for arc propagation and the resulting speed of arc development. These requirements and behaviors of arc propagation will be further discussed in detail. Wilkins previously expressed the relationship between the arc radius, arc current, and arc density as follows:

$$r_{\text{arc}} = \sqrt{\frac{I_{\text{arc}}}{A\pi}} \tag{9.7}$$

where I_{arc} represents the arc current and A is the current density. Obviously, the current density distribution along the surface between arc and water band is not uniform. Assume the current density's maximum value around the arc root $J_{\text{arc-max}}$ is:

$$J_{\text{arc-max}} = \frac{k_{\max} I_{\text{arc}}}{\pi r_{\text{arc}}} \tag{9.8}$$

where $k_{\max}$ is the distribution factor of the surface current near the arc roots. Therefore, the max electric field near the arc root is:

$$E_{\text{arc-max}} = \frac{J_{\text{arc-max}}}{\Delta h_{\text{band}} \sigma_{\text{water}}} = \frac{k_{\max} I_{\text{arc}}}{\pi r_{\text{arc}} \sigma_{\text{water}} \Delta h_{\text{band}}} \tag{9.9}$$

where σ_{water} and Δh_{band} are the water conductivity and height of the water band. It is obvious that the arc root cannot remain stable when $E_{\text{arc-max}}$ is larger than the air breakdown electric field E_c. Instead, the arc root will move forward, or enlarge its radium. Therefore, the flashover requirement of "dry band-water band" on the hydrophobic surface is met when $E_{\text{arc-max}}$ equals E_c and the following equation holds:

$$I_{\text{arc}} = \frac{\pi E_c \sigma_{\text{water}} \Delta h_{\text{band}}}{k_{\max}} r_{\text{arc}} \tag{9.10}$$

Define $I_{\text{arc-c}}$ as the critical current value that arc root can keep developing. When the arc current is larger than $I_{\text{arc-c}}$, the electric field around the arc root is larger than E_c, which would lead to the growth of arc. In addition, the decrease of water band resistance would cause the increase of the arc current, eventually leading to the flashover of the "dry band-water band" surface.

Substituting Eq. (9.7) into Eq. (9.10) yields:

$$I_{\text{arc-c}} = \frac{\pi \sigma_{\text{water}}^2 E_c^2 \Delta h_{\text{band}}^2}{A k_{\max}^2} \tag{9.11}$$

Hence, the prerequisites of the flashover on the hydrophobic surface are listed below:

1. The field intensification is strong enough for the occurrence of dry band breakdown.

2. The arc current is larger than I_{arc-c}, which is linear to the squared water conductivity of the water band.

9.5 Summary

In this chapter, the flashover processes involving water droplets and water bands on hydrophobic surfaces were experimentally investigated, with particular attention given to how varying water parameters influence flashover behavior. The main conclusions drawn from this study are as follows:

1. The flashover voltage is closely related to the length of the water droplet. Specifically, the flashover voltage with water droplets present is similar to the flashover voltage of a dry hydrophobic surface and slightly lower than the pure dry-air flashover voltage.
2. Water conductivity does not significantly affect the flashover voltage, but it significantly alters the arc path. At high conductivity, the arc connects directly through the droplet, forming a series connection, whereas at low conductivity, the arc bridges the droplet, forming a parallel connection.
3. When the test voltage exceeds the dry band breakdown voltage, a partial arc forms. If the water band is shorter than the CFLWB, the partial arc propagates along the water band, resulting in flashover. However, if the water band length exceeds the CFLWB, the partial arc remains stationary, ultimately evaporating or destroying the water band without flashover.
4. The CFLWB increases linearly with increasing water conductivity and dry band length. Moreover, as the width of the water band increases, the CFLWB initially rises before stabilizing.
5. Flashover occurrence on a hydrophobic surface with a water band requires two essential conditions: first, the applied voltage must exceed the dry band breakdown voltage; second, the arc current must exceed a critical value (I_{arc}), which scales linearly with the square of the water band's conductivity.

Bibliography

Bojovschi, A., Rowe, W., and Wong, K. The influence of hanging water droplets on discharge activity, application to high voltage insulators, Appl. Phys. Lett., 2011, 98 (9), 091504.

Bruggeman, P., Guns, P., Degroote, J. et al. Influence of the water surface on the glow-to-spark transition in a metal-pin-to-water electrode system, Plasma Sources Sci. Technol., 2008, 17 (4), 045014.

Cao, B., Shao, S., Wu, X. et al. Flashover characteristics of a hydrophobic surface covered by water, J. Phys. D: Appl. Phys., 2021, 54 (7), 075501.

Fujii, O., Honsali, K., Mizuno, Y. et al. A basic study on the effect of voltage stress on a water droplet on a silicone rubber surface, IEEE Trans. Dielectr. Electr. Insul., 2009, 16 (1), 116–122.

Fujii, O., Honsali, K., Mizuno, Y. et al. Vibration of a water droplet on a polymeric insulating material subjected to AC voltage stress, IEEE Trans. Dielectr. Electr. Insul., 2010, 17 (2), 566–571.

Guan, Z. Composite insulator technologies, In External Insulation of Insulators and Power Transmission, Transformation Equipments, 1st ed., 2006, pp. 268–269.

Guan, Z., Yu, S., and Jia, Z. The pollution and hydrophobicity properties of RTV coating, High Volt., 1993, 19 (4), 28–32.

Hampton, B.F. Flashover mechanism of polluted insulation, Proc. Inst. Electr. Eng., 1964, 111, 985–990.

He, L. and Gorur, R.S. Source strength impact analysis on polymer insulator flashover under contaminated conditions and a comparison with porcelain, IEEE Trans. Dielectr. Electr. Insul., 2016, 23 (4), 2189–2195.

Jiang, X.L., Zhao, S.H., Xie, Y.B. et al. Study on fog flashover performance and fog-water conductivity correction coefficient for polluted insulators, IET Gener. Transm. Dis., 2013, 7 (2), 145–153.

Jolly, D.C. Contamination flashover, Part I: theoretical aspects, IEEE Trans. Power Appar. Syst., 1972, PAS-91 (6), 2437–2442.

Liang, X. and Li, S. Looking to the future of composite insulators, 2015 INMR World Congress, Munich, Germany, 2015, pp. 18–21.

Liang, X., Gao, Y., Wang, J. et al. Rapid development of silicone rubber composite insulator in China, High Volt., 2016, 42 (9), 2888–2896.

Liu, Y. and Du, B.X. Energy eigenvector analysis of surface discharges for evaluating the performance of polymer insulator in presence of water droplets, IEEE Trans. Dielectr. Electr. Insul., 2014, 21 (6), 2438–2447.

Liu, Y., Du, B., and Farzaneh, M. Self-normalizing multivariate analysis of polymer insulator leakage current under severe fog conditions, IEEE Trans. Power Del., 2017, 32 (3), 1279–1286.

Liu, Y., Wu, Y., and Du, B.X. Dynamic formation mechanism of water droplet and induced surface discharges on SiR composites, High Volt., 2019, 4 (1), 59–64.

Nacke, H. Stability of foreign layer discharges and theory of contamination flashover, ETZ-A, 1966, 87, 577–585.

Nazemi, M.H. and Hinrichsen, V. Experimental investigations on water droplet oscillation and partial discharge inception voltage on polymeric insulating surfaces under the influence of AC electric field stress, IEEE Trans. Dielectr. Electr. Insul., 2013, 20 (2), 443–453.

Obenaus, F. Contamination flashover and creepage path length, Deutsche Elektrotechnik, 1958, 12, 135–136.

Ouis, A. and Flazi, S. DC flashover criterion and discharge elongation mechanism in an air–electrolyte interval, IET Gener. Transm. Dis., 2015, 9 (1), 53–60.

Salem, A.A., Rahman, R.A., Kamarudin, M.S. et al. Factors and models of pollution flashover on high voltage outdoor insulators: review, IEEE Conference on Energy Conversion (CENCON), 2017, pp. 241–246.

Shkuropat, P.I. Development of a discharge on a wet insulator surface with d.c, N-Tekh. Inf. Byull. Leningrad Polytechnic Inst, 1957, 41–51.

Soares, D., Mendonça, S., Neto, E. et al. Electrical field on non-ceramic insulators and its relation to contact angles for constant volume droplets, J. Electrost., 2016, 84, 97–105.

Wilkins, R. and Al-Baghdadi, A.A.J. Arc propagation along an electrolyte surface, Proc. Inst. Electr. Eng., 1971, 118 (12), 1886–1892.

Wilkins, R. Flashover voltage of high-voltage insulators with uniform surface-pollution films, Proc. Inst. Electr. Eng., 1969, 116 (3), 457–465.

Yang, L. Evaluation of wetting condition and its effects on pollution flashover voltage of aerodynamic insulators, IEEE Trans. Dielectr. Electr. Insul., 2016, 23 (5), 2875–2882.

Zhang, C.Y., Hu, J.Y., Li, J.C. et al. Experimental study on the contamination deposition characteristics of insulators in a fog–haze environment, IET Gener. Transm. Dis., 2018, 12 (2), 406–413.

Zhang, C.Y., Wang, L.M., Guan, Z.C. et al. Pollution flashover performance of full-scale ±800 kV converter station post insulators at high altitude area, IEEE Trans. Dielectr. Electr. Insul., 2013, 20 (3), 717–726.

Zhu, Y., Otsubo, M., Honda, C. et al. Loss and recovery in hydrophobicity of silicone rubber exposed to corona discharge, Polym. Degrad. Stab., 2006, 91 (7), 1448–1454.

10

Factors Influencing the Hydrophobic Surface Flashover and Their Application

10.1 Introduction

Traditionally, it has been assumed that the presence of isolated water droplets—as opposed to continuous water bands or films—on insulator surfaces poses minimal risk, and such conditions are generally considered "safe," implying flashover is unlikely to occur. However, operational experiences and recent studies have shown instances where flashover occurred on insulator surfaces that maintained high hydrophobicity and contained only discrete water droplets. Clearly, while hydrophobicity significantly reduces or prevents surface flashover under typical conditions, certain specific scenarios can still lead to flashover incidents. In Chapter 9, the necessary conditions for flashover on hydrophobic surfaces were identified. This chapter builds upon that foundation by analyzing the flashover threshold from the initial onset of wetting to complete surface saturation, and by examining the key factors influencing hydrophobic surface flashover. This analysis offers valuable insight into the behavior of flashover voltage under wet conditions and provides meaningful guidance for practical engineering applications.

10.2 Wetting Prerequisite of Flashover on Hydrophobic Surface

For insulators with strong hydrophobicity, moisture on the surface typically forms discrete water droplets rather than continuous films. It is well known that the breakdown electric field strength in a uniform air field is approximately 30 kV/cm. Partial discharges, such as corona or partial arcs, occur only when the local electric field strength exceeds this threshold.

However, because the electric field distribution along an insulator surface is inherently nonuniform, the breakdown threshold under uneven field conditions is much lower, typically around 4–5 kV/cm (average electric field strength). Therefore, for a flashover to occur under standard operating voltage levels, a significant distortion and intensification of the local electric field must take place.

The presence of water droplets or water bands causes significant local field enhancement, as illustrated in Figure 10.1, particularly around the edges of the droplets or bands. This local intensification creates favorable conditions for initiating partial discharges, which can lead to flashover under certain conditions.

When water droplets or bands are present on a hydrophobic surface, they must be sufficiently large or elongated to focus the electric stress on the adjacent dry band above its breakdown threshold. Once that threshold is exceeded, the arc propagates along the surface, traversing the water band—a process that demands more energy than initiation alone. As derived in Chapter 9, there exists a critical arc current required for sustained propagation: if the actual current exceeds this critical value, the arc will bridge the water band and cause a full flashover; if it does not, the arc will remain localized or eventually extinguish.

In Chapter 9, we observed that water droplets consistently reduce flashover voltage, regardless of their volume. At first glance, this seems to contradict the results of full-scale pollution flashover tests on composite insulators, which show that early-stage wetting does not lower flashover voltage. However, in Section 9.3 the droplets ranged from 0.5 cm to 4.0 cm in length within a total leakage distance of only 5.0 cm—meaning they occupied a substantial portion of the insulation path. This likely explains the voltage reduction we measured, suggesting the effect is strongly size-dependent: large droplets significantly shorten the effective gap and thus lower the flashover voltage.

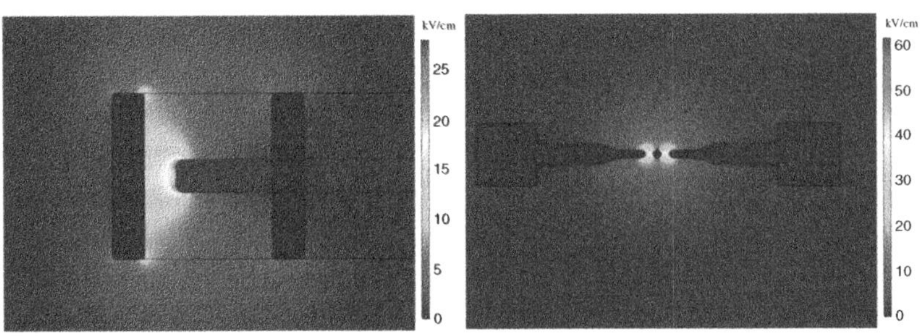

(a) (b)

Figure 10.1 Electric field intensification caused by water: (a) water band; (b) water droplet.

To investigate this hypothesis, a supplementary experiment was conducted focusing on surface flashover with smaller water droplets.

For this set of experiments, the test circuit remained the same as that described in Section 9.2.1. A polytetrafluoroethylene (PTFE) board was used as the test specimen, with a defined test area of 10.0 cm × 5.0 cm. Conductive tape was attached to both ends of the board to serve as electrodes, and the insulation length was fixed at 5.0 cm. Two acrylic support cubes were used to elevate the specimen and prevent flashover along the backside.

During the test, water droplets of varying small sizes, all with conductivity less than 10.0 μS/cm, were carefully placed in the center of the PTFE surface. The voltage was gradually increased until flashover occurred. The same high-speed camera and DAQ system used in Chapter 2 were employed for observation and data acquisition. The actual test setup is shown in Figure 10.2.

The length of the water droplet just before flashover varies from 1.0 mm to 20.0 mm and the flashover voltage is recorded as well. The results are shown in Figure 10.3.

The length of each water droplet was measured immediately before flashover, varying from 1.0 mm to 20.0 mm, and the corresponding flashover voltage was recorded. The results are plotted in Figure 10.3: for droplet lengths ≤10.0 mm, there is no measurable effect on flashover voltage. In this range, the flashover voltage remains at approximately 34.9 kV (peak), essentially identical to the value measured on a dry surface. While for droplet lengths >10.0 mm, the flashover voltage decreases linearly as droplet length increases. This trend closely mirrors the behavior observed in air-gap breakdown tests, where the effective gap reduction determines the breakdown voltage. These findings confirm a size-dependent threshold: only droplets occupying more than about 20% of the 50 mm insulation path significantly reduce flashover voltage.

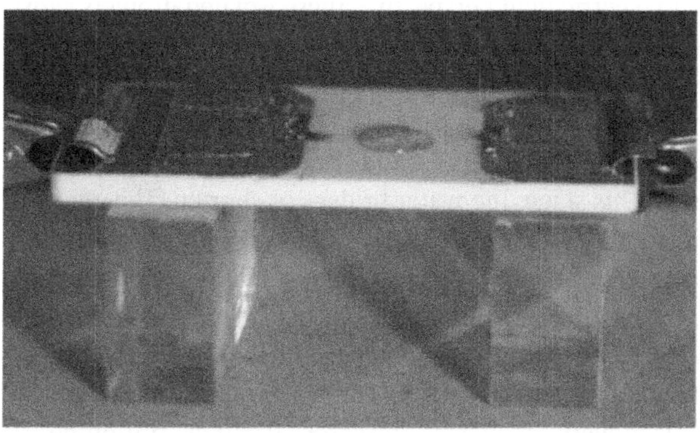

Figure 10.2 Sample arrangement.

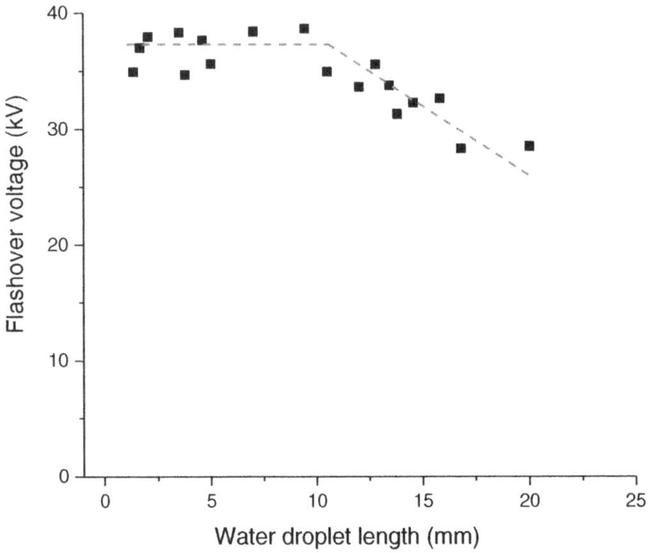

Figure 10.3 Relationship between flashover voltage and water droplet diameter.

10.3 Effect of Contamination on Flashover Voltage Under Different Wetting Conditions

As the experimental results indicate, the flashover behavior of wetted hydrophobic surfaces can be categorized into two primary stages, based on how water coverage influences flashover voltage:

Stage 1 (Small Water Proportion): When the proportion of surface covered by water is small, the presence of water droplets does not significantly affect the flashover voltage. From an electric field perspective, these isolated droplets cause minimal local field distortion, thus insufficient to initiate flashover. Consequently, the flashover voltage remains close to that of a completely dry surface.

Stage 2 (Large Water Proportion): When the water coverage increases beyond a critical proportion, water droplets or water bands begin to actively participate in the flashover process. At this stage, the water bridges or shortens the effective insulation path, creating conditions favorable for arc initiation and propagation. The flashover voltage then decreases linearly with the residual insulation length, as clearly demonstrated by experimental observations.

Traditionally, it is widely believed that composite insulators, characterized by excellent hydrophobicity and hydrophobicity transfer properties, are largely resistant to pollution flashovers. However, research findings presented in Section 9.3 highlight that even limited amounts of water can lead to measurable reductions in flashover voltage, prompting a need to reassess the relationship between flashover voltage, contamination degree, and wetting level, as will be discussed subsequently.

Pollution flashover under working conditions remains a critical concern due to its potential impact, including difficulty in system reclosing or even widespread blackouts. Consequently, transmission line inspection programs routinely assess the hydrophobicity of insulator surfaces. Typically, power utilities adopt a cautious approach by replacing composite insulators once hydrophobicity degradation is observed. In many instances, aged insulators lose surface hydrophobicity entirely, as their internal low-molecular-weight components become depleted and can no longer migrate to the surface to restore hydrophobicity. Replacing such aged insulators is necessary, given their substantially deteriorated insulation performance. However, insulators still possessing acceptable hydrophobicity levels retain significantly higher flashover voltages compared to fully aged composite or conventional inorganic insulators. Thus, replacing insulators based solely on minor hydrophobicity reduction may result in unnecessary expenditures of time and resources.

Considering these factors, it is critical to precisely understand the relationship among flashover voltage, contamination degree, and wetting degree. Based on the results of this research, the wetting process affecting insulation performance can be further detailed into three distinct stages:

1. Initial Wetting Stage: From the onset of wetting until droplets begin affecting the flashover voltage, droplets remain small and isolated. In this stage, flashover voltage remains nearly equal to that of a dry insulator surface.
2. Intermediate Wetting Stage: As wetting progresses, droplets grow or merge to form water bands, initiating measurable influence on flashover voltage. During this stage, the flashover voltage decreases linearly with the length of the residual dry insulation gap. Notably, water conductivity has minimal effect on flashover voltage at this stage.
3. Saturated Wetting Stage: Eventually, insulators become extensively covered by continuous water films or dense water bands, forming dominant conductive pathways. In this stage, flashover voltage is primarily determined by surface resistance, directly influenced by the contamination level and conductivity of the water.

Understanding and distinguishing these stages can significantly improve maintenance strategies, enabling utilities to better assess when insulators require immediate replacement or cleaning, thereby optimizing resource allocation and enhancing operational reliability. The above flashover voltage trends are visualized in Figure 10.4.

In engineering practice, it is challenging for significant amounts of water to accumulate on composite insulator surfaces that retain strong hydrophobicity. This difficulty arises due to the relatively small contact angle of water droplets, as well as external factors such as gravity, wind drag, and mechanical vibration, all of which tend to dislodge droplets from the surface. Consequently, most operational

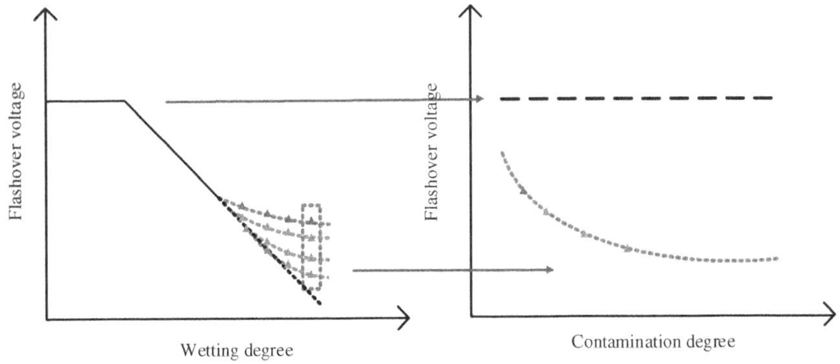

Figure 10.4 The trend diagram for relationship between flashover voltage contamination degree and wetting degree.

conditions align with Stage 1, where the proportion of water on the insulator surface is small, resulting in minimal impact on the flashover voltage. This behavior underpins the widespread belief within the power industry that composite insulators inherently resist pollution flashover.

However, under certain extreme or special circumstances, substantial amounts of water can accumulate on the hydrophobic surface. In these cases, categorized as Stage 2, the increased proportion of water significantly intensifies local electric fields and effectively reduces the insulation path length. Consequently, this leads to a notable decrease in flashover voltage.

In Stage 3, where hydrophobicity is largely compromised or entirely lost, the flashover voltage becomes strongly dependent on the conductivity of the water film, which itself is governed by contamination levels. This relationship has been consistently validated by artificial pollution experiments and various studies.

An important phenomenon observed during the transition from Stage 2 to Stage 3 is the vibration-induced coalescence of water droplets. Mechanical vibrations facilitate the merging of droplets, encouraging the formation of larger droplets or continuous water bands. This process and its implications on flashover performance are discussed in detail in the following section.

10.4 Dynamic Behavior of Water Droplets Under Electric Field

Unlike the flashover on the hydrophilic surface, it is commonly accepted that the flashover mechanism on the water droplets covered hydrophobic surface can be summarized as follows: with the increase of the electric field, water droplets are first elongated and then be merged to form water bands, after that the air gaps between water bands are shortened and broken down, leading to the occurrence of flashover. For composite insulators, the water will form individual water droplets on the insulator surface with a relatively high contact angle. In other words, the water droplets are close to the ideal sphere so it can keep enough insulation length. Once the water droplets are elongated and merged, the residual insulation length is shortened which may give a chance to flashover. It is also vital to study the dynamics of water droplets on the hydrophobic surface.

The flashover voltage is highly related to the residual insulation length on the hydrophobic surface as mentioned. The water proportion is negative linear to the flashover voltage. In terms of Chapter 2, the water band is fixed on a PTFE board with AC electric field so the water band cannot be elongated. In Chapter 3, the water droplet passes through the air gap within less than 0.01 s under DC electric field. The deformation of water droplet is observed in Chapter 3 indeed. But the dynamic is not complete. Under actual outdoor operation conditions, the water droplet on the insulator surface may be elongated and merged under AC voltage during wetting. In the presence of the electric field, the water droplets are subject mainly to the Maxwell stress tensor and Coulomb force. Under a uniform AC electric field, the Maxwell stress tensor causes the water droplets to oscillate symmetrically with a frequency two times the applied voltage frequency, while the Coulomb force makes the water droplets shift leftward and rightward periodically with the same frequency as that of the applied voltage. Therefore, to thoroughly understand the flashover process on the hydrophobic surface, it is of great importance to study the dynamics of water droplets sitting on the hydrophobic surface just before flashover. Based on the above consideration, this section will mainly focus on the asymmetric deformation of water droplets under AC electric field.

In this section, the dynamics of the water droplet placed on the hydrophobic surface before flashover was observed using a high-speed camera. The high voltage circuit is almost the same as in Chapter 2. Two round copper plates are used as electrodes, while a silicone rubber sheet with a water droplet on its center is placed in the middle of the copper plate electrodes. The radius of copper plate electrodes was 50 mm and the distance between electrodes was 14.8 mm. The water droplet's existence will intensify the electric field resulting in a nonuniform field. The corona discharge may happen before flashover occurs. Thus, a small DC charge test is also prepared. A surface potentiometer (Trek 341B) is used to measure the surface potential above the water droplet which can indicate the charge inside of water droplet. The test platform for vibration and the surface potentiometer is shown in Figures 10.5 and 10.6.

As charge measurement, first, a 0.02 μL deionized water droplet (<10 μS/cm) was placed on the middle of the HTV silicone rubber board. Then, the HV needle electrode was fixed above the water droplet with 12 kV/cm DC voltage applied. The voltage stress lasted for five minutes and then the potential above the water droplet is measured by the surface potentiometer.

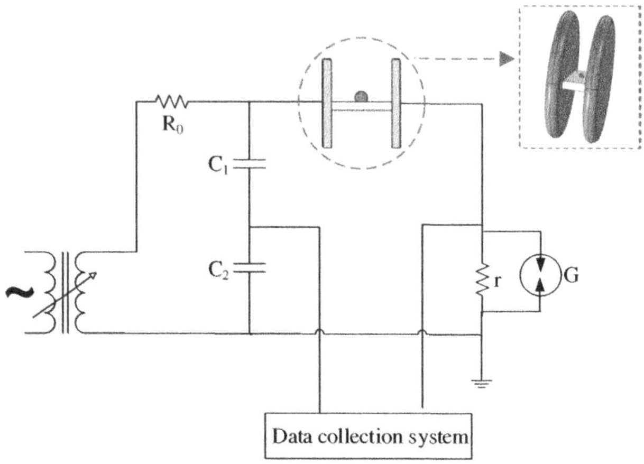

Figure 10.5 Test platform.

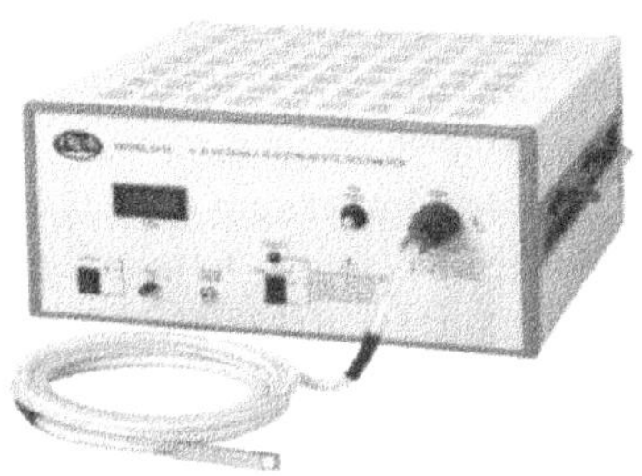

Figure 10.6 Surface potentiometer.

For the water droplet dynamic, at first, water droplets with different water conductivity were prepared using sodium chloride. The water droplet was deposited in the middle of the specimen surface by a syringe. Then, the AC voltage was applied between electrodes and increased at the speed of 1.4 kV/s. The deformation and oscillation under the electric field were captured by means of the high-speed camera. This test was repeated 60 times to find out the distribution of the charge amounts inside water droplets. The result is shown in Figure 10.7.

As it can be seen, water droplets are charged with both positive charge and negative charge. The number of water droplets with different charge polarities is also close. Furthermore, the water droplet's potential distribution is closed to the normal distribution.

However, for the convenience of experiments, the charge measurement test was done under DC voltage. The charging characteristic may be different under AC voltage which can be further studied in the future.

As for the water droplet dynamics, the water droplet will start oscillating under AC electric field. Figure 10.8 shows the shape of water droplet at different

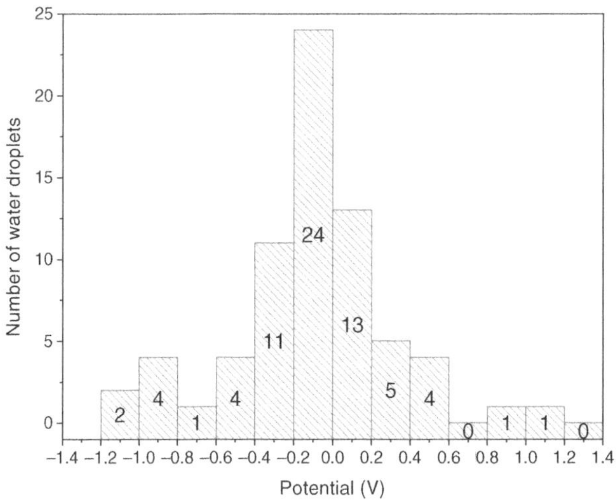

Figure 10.7 Potential distribution of 60 charged water droplets.

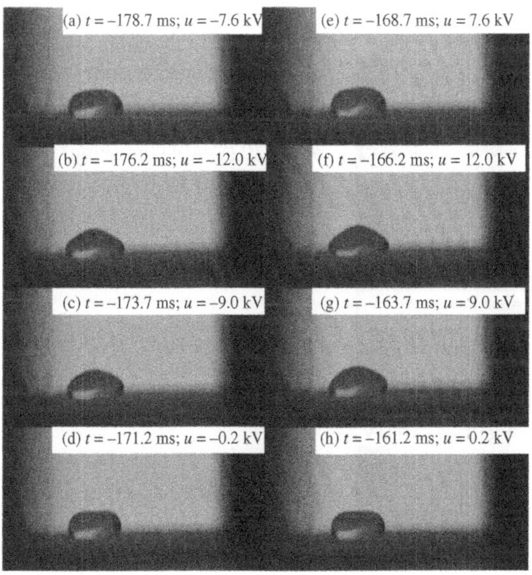

Figure 10.8 Dynamic behavior of the water droplet under AC voltage during the 9th cycle before the flashover. *Source*: Yue et al. (2025)/AIP Publishing/CC BY 4.0.

time points during the test and Figure 10.9 shows the waveform of applied voltage at different time points close to the flashover (the 9th cycle before the flashover). It can be found that the water droplet vibration frequency was two times the applied voltage frequency, namely 100 Hz. With AC voltage stress, the

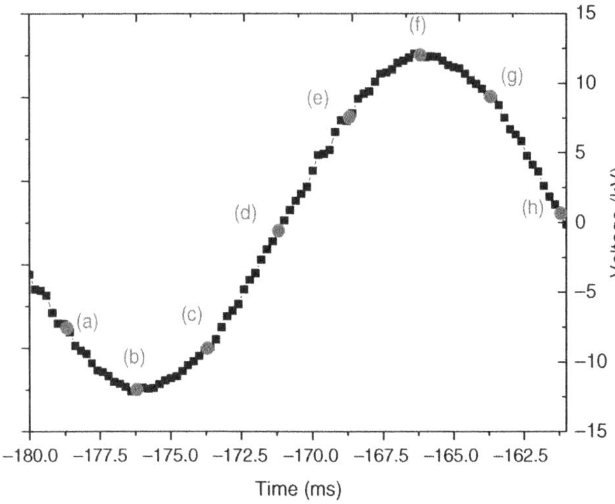

Figure 10.9 The applied voltage with time during the 9th cycle before the flashover (−180 to −160 ms). *Source*: Yue et al. (2025)/AIP Publishing/CC BY 4.0.

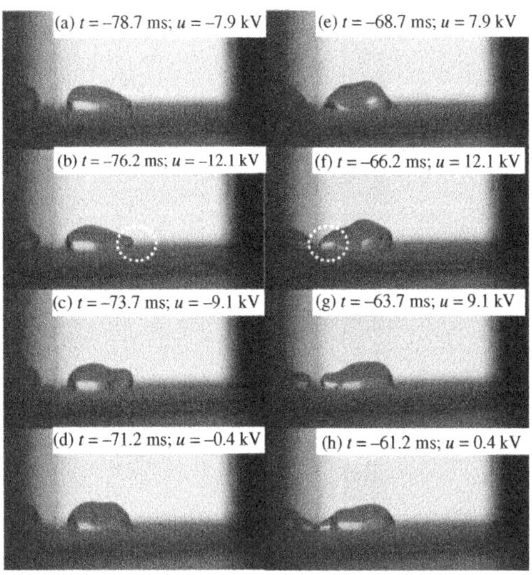

Figure 10.10 Dynamic behavior of the water droplet under AC voltage during the 4th cycle before the flashover. *Source*: Yue et al. (2025)/AIP Publishing/CC BY 4.0.

water droplet was elongated periodically along the electric field direction and got the maximum when the applied voltage reached its peak value as shown in Figure 10.8b,f. During the 9th cycle before the flashover, the shape of the water droplet was approximately symmetrical.

Figures 10.10 and 10.11 describe the water droplet shape and voltage waveform in the 4th cycle before flashover. It can be found that the water droplet remained

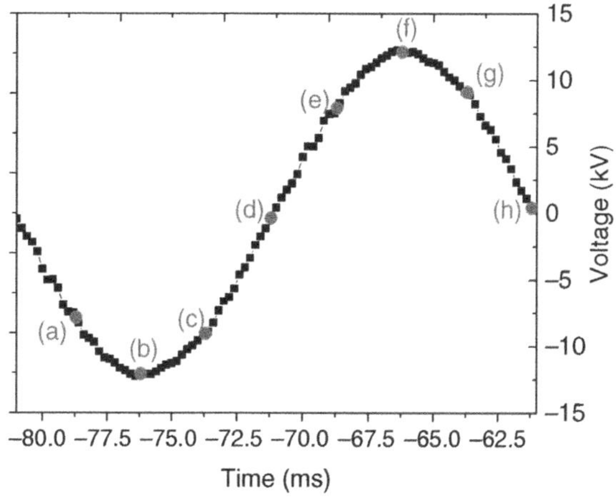

Figure 10.11 The applied voltage with time during the 4th cycle before the flashover (−80 to −60 ms). *Source:* Yue et al. (2025)/ AIP Publishing/CC BY 4.0.

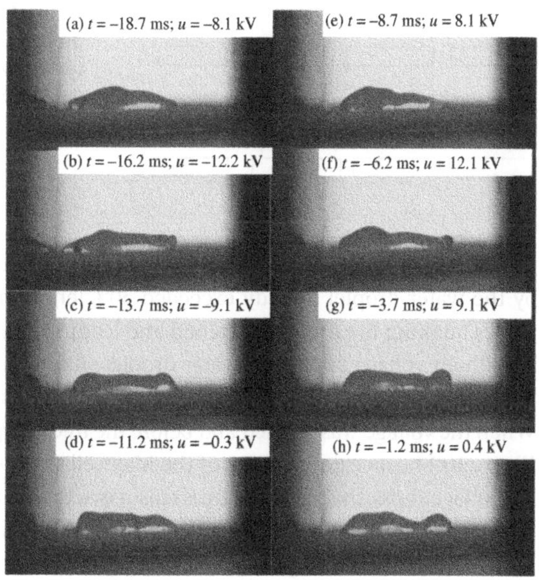

Figure 10.12 Dynamic behavior of the water droplet under AC voltage during the cycle just before the flashover. *Source:* Yue et al. (2025)/AIP Publishing/ CC BY 4.0.

most stretched when the voltage was at its peak. Unlike the phenomenon observed in Figure 10.8, the shape of water droplet was no longer symmetrical during the elongation process. In the negative half cycle, the area where the droplets protruded, circled by dash lines as indicated in Figure 10.10b, was on the right side. However, it changed alternately with the polarity of applied voltage as shown in Figure 10.10f.

Figures 10.12 and 10.13 above illustrate the shape of the water droplet and applied voltage during the cycle just before the flashover. As can be seen from

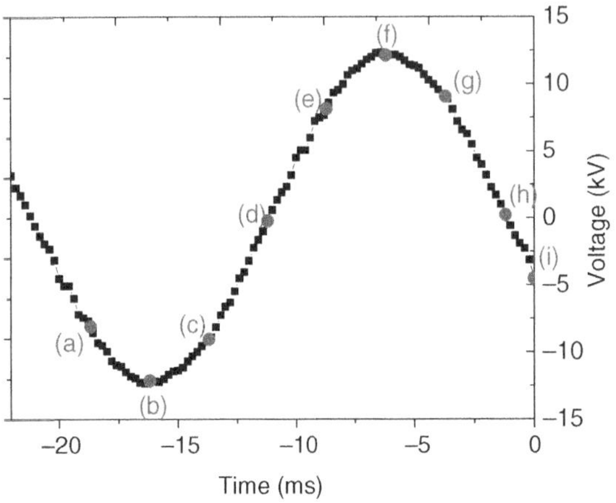

Figure 10.13 The applied voltage with time just before the critical flashover (−20 to 0 ms). *Source*: Yue et al. (2025)/AIP Publishing/ CC BY 4.0.

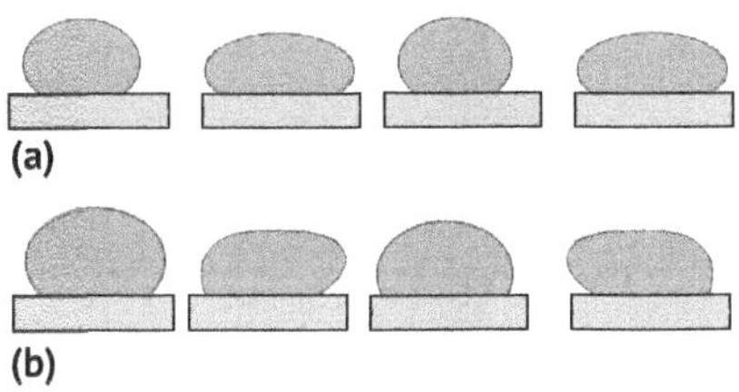

Figure 10.14 Deformation of water droplet under AC electric field: (a) without free charge; (b) with free charge. *Source*: Yue et al. (2025)/AIP Publishing/CC BY 4.0.

Figure 10.12, the water droplet was elongated dramatically and the area occupied by the water droplet already exceeded 50% length of the gap between two electrodes, making the droplet flattened and leading to a flashover eventually.

Without voltage applied, a water droplet sitting on a horizontal placed uniform hydrophobic surface was axisymmetric due to the symmetry of surface tension. When the voltage was applied and before the free charge was produced, as shown in Figure 10.14a, an external force of the Maxwell stress tensor was introduced. Under AC voltage stress, the Maxwell stress tensor is acting on the water droplet and is highly related to the water droplet shape. As the shape of the water droplet and the Maxwell stress tensor force acting on the water droplet are both bilateral symmetrical, the water droplet vibrated symmetrically before the generation of free charges as shown in Figure 10.14a. However, when the local electric field at the water droplet surface exceeded the corona onset strength, free charges were generated by the corona discharge. And the polarity of the charge is random according to the charge test. In this condition, the water droplet was also affected by the Coulomb force in addition to the Maxwell stress tensor. The Coulomb force acting on the water droplet was not symmetrical and the force direction depended on the polarity of free charge and the direction of the electric field. Therefore, the water droplet was deflected to one side and the other side alternatively within each cycle as shown in Figure 10.14b. For superhydrophobic coated surface, the water droplet can easily roll off on its surface

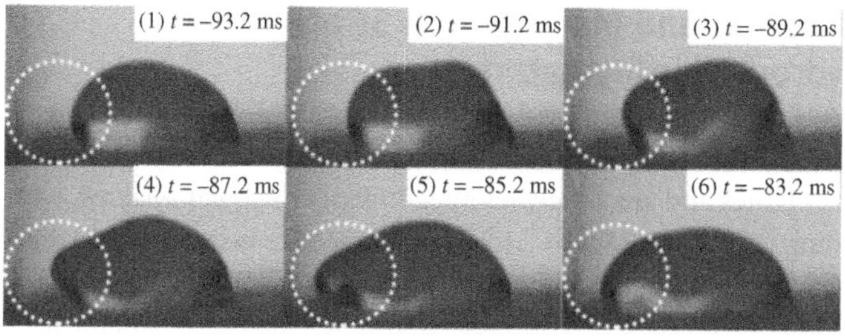

Figure 10.15 The water droplet elongation process during 93.2–83.2 ms just before the critical flashover. *Source*: Yue et al. (2025)/AIP Publishing/CC BY 4.0.

due to a small rolling angle and the flashover voltage is rather high. On the contrary, for silicone rubber surfaces without superhydrophobic coating, the water droplet will cover a larger area of the silicone rubber surface due to the Coulomb force as shown in Figure 10.15 and finally lead to a lower flashover voltage.

When the applied voltage changes with time, the polarization relaxation time of the cloud droplet can be derived as follows:

$$\nabla \cdot \boldsymbol{J} + \frac{\partial \rho_{de}}{\partial t} = 0 \tag{10.1}$$

$$\boldsymbol{J} = \sigma_{water} \boldsymbol{E} \tag{10.2}$$

Equations (10.1) and (10.2) are the charge continuity and charge current equations, where J and ρ_{de} represent the current density and charge density in the cloud droplet, σ_{water} is the conductivity of water. According to Poisson's equation

$$\nabla \cdot \boldsymbol{E} = \frac{\rho_{de}}{\varepsilon_0 \varepsilon_{water}} \tag{10.3}$$

where ε_0, ε_{water} represent vacuum permittivity and relative permittivity of water, substituting Eqs. (10.2) and (10.3) into Eq. (10.1) yields

$$\sigma_{water} \frac{\rho_{de}}{\varepsilon_0 \cdot \varepsilon_{water}} + \frac{\partial \rho_{de}}{\partial t} = 0 \tag{10.4}$$

Thus, the charge density is proportional to $\exp(-\sigma t/(\varepsilon_0 \varepsilon_{water})$ and the time constant $\varepsilon_0 \varepsilon_{water}/\sigma_{water}$ is the polarization relaxation time. For distilled water droplet with a conductivity of 0.5 μS/cm, the polarization release time is 1.6×10^{-5} seconds. As the above conductivities are much higher than 0.5 μS/cm, the polarization relaxation time of the droplet can be ignored and the charge redistribution is assumed to be obtained instantly during the change of applied voltage. Therefore, the water conductivity has no obvious effect on the deformation of the water droplet.

10.5 Replacement and Maintenance Strategies for Degraded Hydrophobic Insulators

As discussed previously, composite insulators have significantly alleviated pollution flashover issues and are widely adopted in modern power systems, with their use continuing to increase rapidly. However, composite insulators possess a significant drawback in terms of aging. Due to their organic material properties, composite insulators have inherently limited lifetimes. When an insulator ages beyond a certain point, its hydrophobicity gradually decreases and may even be completely lost in severe cases. For safety and reliability considerations, aged or deteriorated insulators must then be replaced promptly.

Nevertheless, the aging process of composite insulators is complex, involving multiple dimensions such as hydrophobicity loss, surface hardness reduction, color changes, diminished mechanical strength, and surface pulverization. Although extensive research into aging mechanisms has been continuously conducted, no definitive and universally accepted conclusions have been reached. Additionally, there currently exists no standardized criterion for quantifying insulator aging, resulting in ambiguity when determining replacement thresholds, particularly concerning reduced hydrophobicity.

In practice, criteria for insulator replacement during routine inspections primarily depend on the subjective judgment and experience of field inspectors. Insulators exhibiting critical defects are promptly replaced. However, even in cases where defects are minor or limited to occasional partial discharges, operating departments often adopt a conservative approach, choosing replacement to preclude potential accidents. This conservative practice results in the unnecessary replacement of a significant number of insulators annually.

According to the research presented in Chapter 9, if hydrophobicity is not completely lost, it can still significantly enhance flashover voltage. Some researchers involving artificial pollution flashover experiments at varying hydrophobicity classification (HC) levels have demonstrated that even minimal residual hydrophobicity can notably improve flashover voltage. This suggests that among the insulators currently being replaced, a substantial proportion retain sufficient hydrophobicity and functional capability despite minor partial discharges caused by heavy contamination, raindrops, or other extreme environmental conditions. Such incidents typically represent low-probability events that rarely escalate into complete pollution flashovers.

To prevent considerable economic losses and minimize service outages due to premature insulator replacements, it is crucial to identify and define a scientifically sound replacement threshold based on quantified assessments of hydrophobicity and its direct impact on flashover performance.

Based on the discussion above, it is evident that the criteria for insulator replacement must be carefully reconsidered. To ensure both safe operation and efficient maintenance, a quantitative and reliable indicator is essential for evaluating the condition of composite insulators. Additionally, the method of evaluation must be practical and feasible for on-site application.

Among various diagnostic approaches, the measurement of surface hydrophobicity remains the most practical option under real-world conditions. First, hydrophobicity assessment—particularly using the HC method—can be conducted directly in the field without the need to remove the insulator for laboratory testing. Second, surface hydrophobicity is directly related to wetting behavior and the residual insulation path, both of which influence flashover voltage to a significant extent.

According to the Chinese standard DL/T 864-2004, insulators exhibiting hydrophobicity classifications below HC5 should be subject to further evaluation and may need to be withdrawn from service if performance concerns are confirmed. In addition, during periods of high humidity and pollution, such as in summer, insulators within the HC5 to HC7 range are considered to have elevated flashover risk and are often recommended for preventive replacement. These practices reflect long-standing industry experience linking surface hydrophobicity with insulation reliability under wet and contaminated conditions. As such, hydrophobicity classification remains a practical and widely used metric for assessing the condition of composite insulators in the field.

As demonstrated in the previous analysis, it is only when the proportion of water on the insulator surface becomes non-negligible that the insulation path is significantly curtailed, leading to a reduction in flashover voltage. Therefore, in practical maintenance and operation, attention should primarily be given to cases where hydrophobicity is insufficient.

Based on the established relationship between flashover voltage, contamination level, and wetting degree, the criteria for assessing whether an insulator should be withdrawn from service can be further refined. This allows maintenance decisions to be more targeted, reducing unnecessary replacements while maintaining safety. Taking both economic efficiency and insulation reliability into account, the following recommendations are proposed for managing insulators with reduced hydrophobicity:

HC1 to HC4: Insulators in this range are considered to have good hydrophobicity and may continue operating without any maintenance.

HC5: Insulators may remain in service but should be subject to ongoing monitoring to detect any deterioration.

HC6 to HC7: These insulators should be removed from service and sent for laboratory evaluation. The hydrophobicity should first be tested with the pollution layer intact. If the HC level remains at HC6 or HC7, the pollution layer should be washed off, and the surface retested.

If hydrophobicity recovers after cleaning, the insulator may be treated (e.g., by applying RTV coating or using insulator detergent) and possibly returned to service.

If hydrophobicity does not improve, the insulator should be permanently replaced.

This classification and assessment strategy has been formally adopted in the China power industry standard DL/T 1474-2021, providing a more nuanced and economically sound approach to insulator maintenance while ensuring continued operational safety.

10.6 Summary

This chapter reviewed the discharge characteristics of composite insulators in the presence of water droplets, synthesizing key findings from the preceding chapters. It outlined the fundamental requirements for flashover in composite insulators and examined two critical factors influencing the flashover process: wetting degree and water dynamics. The insights gained were then applied to propose a practical framework for evaluating and replacing composite insulators with reduced hydrophobicity. The main conclusions are summarized as follows:

1. The relationship between flashover voltage, contamination degree and wetting degree is: first, the flashover voltage is the same as dry case when wetting degree is low; then the flashover voltage is linear to wetting degree (residual insulation length); at last, when the most area is covered with water, the flashover voltage is related to the conductivity or contamination degree.
2. The charge in water droplets obeys normal distribution under DC voltage stress, the free charge will change water droplets from symmetrical to asymmetrical oscillations owing to the Coulomb force. Specially, at the instant just before flashover, the water droplet will be flattened that obviously reduces the residual insulation length resulting in reduction of flashover voltage.
3. Through the study of wetting degree and flashover voltage, the criterion for replacement of composite insulators with decreased hydrophobicity is proposed which can reduce unnecessary waste and power cuts for maintenance.

Bibliography

Boudissa, R., Belhoul, T., and Haim, K.D. Effect of inclination angle of hydrophobic silicone insulation covered by water drops on its DC performance, IEEE Trans. Dielectr. Electr. Insul., 2017, 24 (5), 2890–2900.

Cao, B., Shao, S., and Wu, X. Flashover characteristics of a hydrophobic surface covered by water. J. Phys. D: Appl. Phys., 2020, 54 (7), 075501.

Cheng, L., Liu, Y., Chen, R. et al. Method for predicting the water absorption of external insulating silicone rubber. IEEE Trans. Dielectr. Electr. Insul., 2022, 29 (4), 1242–1250.

Chu, F., Wu, X., Zhu, B. et al. Self-propelled droplet behavior during condensation on superhydrophobic surfaces, Appl. Phys. Lett., 2016, 108 (19), 194103.

Dubinov, A.E., Kozhayeva, J.P., Golovanov, V.V. et al. Coalescence of liquid droplets under effect of pulsed-periodic spark discharges, IEEE Trans. Plasma Sci., 2019, 47 (1), 76–80.

Fujii, O., Honsali, K., and Mizuno, Y. A basic study on the effect of voltage stress on a water droplet on a silicone rubber surface, IEEE Trans. Dielectr. Electr. Insul., 2009, 16 (1), 116–122.

Li, J., Wei, Y., Huang, Z. et al. Investigation of the electric field driven self-propelled motion of water droplets on a super-hydrophobic surface, IEEE Trans. Dielectr. Electr. Insul., 2016, 23 (5), 3007–3015. https://doi.org/10.1109/TDEI.2016.7736864.

Li, Y., Jin, H., Nie, S. et al. Effect of super hydrophobicity on flashover characteristics of silicone rubber under wet conditions, AIP Adv., 2018, 8, 15313.

Li, Z., Wang, N., and Dong, K. Flashover characteristics of a droplet impacting the silicone rubber surface, AIP Adv. 2023, 13, 15024.

Naito, K., Izumi, K., Takasu, K. et al. Performance of composite insulators under polluted conditions, CIGRE Paper no. 33-301, 1996.

Nazemi, M.H. and Hinrichsen, V. Experimental investigations on partial discharge characteristics of water droplets on polymeric insulating surfaces at AC, DC and combined AC-DC voltages, IEEE Trans. Dielectr. Electr. Insul., 2015, 22 (4), 2261–2270.

Ndoumbe, J., Beroual, A., and Imano, A.M. Simulation and analysis of coalescence of water droplets on composite insulating surface under DC electric field, IEEE Trans. Dielectr. Electr. Insul., 2015, 22 (5), 2669–2675.

Prasad, D.S. and Reddy, B.S. Study of corona degradation of polymeric insulating samples using high dynamic range imaging technique, IEEE Trans. Dielectr. Electr. Insul., 2017, 24 (2), 1169–1177.

Sarathi, R., Harsha, V.S., Vasa, N.J. et al. Water droplet initiated discharges on epoxy nanocomposites under DC voltages, IEEE Trans. Dielectr. Electr. Insul., 2016, 23 (3), 1743–1752.

Wei, S., Jin, H., Zhou, H. et al. Dynamic behavior of water droplets on wetted superhydrophobic surfaces under a high AC electric field, AIP Adv., 2019, 9, 65307.

Wu, X., Cao, B., Li, Z. et al. DC breakdown characteristic of air gap with water droplets. IEEE Trans. Plasma Sci., 2021, 49 (6), 1962–1968.

Xie, G., He, F., Liu, X. et al. Sessile multidroplets and salt droplets under high tangential electric fields, Sci. Rep., 2016, 6, 25002.

Yue, Y., Qian, B., Cao, B. et al. Influence of free charges on flashover of silicone rubber surface with droplet covered, AIP Adv., 2025, 15, 075037.

Zhang, F., Wang, X., Wang, L. et al. Effect of arcing on DC flashover performance of contaminated porcelain insulators for various suspension patterns at high altitudes, IEEE Trans. Dielectr. Electr. Insul., 2008, 15 (3), 783–791.

Zhang, S., Cheng, L., Liao, R. et al. Process improvement to restrain emergency heating defect of composite insulator. IEEE Trans. Dielectr. Electr. Insul., 2022, 29 (2), 446–453.

11

Insulator Contamination Monitoring Based on Its Hygroscopicity Under Unsaturated Humidity

11.1 Introduction

Building on the theoretical insights into contamination evolution and wetting mechanisms presented in the first 10 chapters, Chapters 11–12 transition to industrial applications by introducing modern techniques for monitoring pollution on insulation surfaces. Special attention is given to real-time monitoring systems that operate under varying humidity conditions, capturing the dynamic behavior of surface contaminants. By linking measured pollution states to predictive maintenance strategies, these chapters provide practical guidance for improving the reliability and operational management of high-voltage insulation systems.

The design and selection of transmission line insulators are fundamentally influenced by the site pollution severity at the installation site. Traditionally, equivalent salt deposit density (ESDD) and non-soluble deposit density (NSDD) have been widely used to characterize insulator contamination due to their simplicity and ease of measurement. However, ESDD measurement requires complete removal and dissolution of surface pollutants in deionized water, making it impractical for real-time, online monitoring. To overcome this limitation, researchers have explored various monitoring techniques, including empirical correlations between ambient air pollutants and insulator contamination, advanced diagnostic methods such as photothermal radiometry, laser-induced breakdown spectroscopy, and infrared imaging, as well as leakage-current-based approaches.

Among these, leakage current monitoring has proven both practically feasible and effective. By analyzing time-domain indicators—such as maximum and effective current, pulse count, and accumulated charge—and frequency-domain characteristics, including harmonic distortions revealed via FFT or wavelet transforms, researchers have developed models capable of assessing flashover risk.

Nonetheless, most existing models are calibrated under saturated humidity or heavy contamination conditions, limiting their effectiveness in typical field scenarios where contamination accumulates under dry conditions and flashovers occur later when humidity rises.

To address this challenge, this chapter introduces a novel approach leveraging the hygroscopic properties of surface contaminants under unsaturated humidity. By examining the multidimensional relationships among environmental parameters (temperature and humidity), leakage current characteristics, and surface contamination levels, a new method for estimating ESDD in real-time is proposed. This methodology enhances early warning capabilities for pollution-induced flashover and supports the digitalization and automation of transmission line monitoring and maintenance.

11.2 Development of Online Monitoring Devices

The general design of a contamination flashover prewarning system can be described in three main stages. First, pollution condition data from key high-contamination areas are continuously collected using an online monitoring system. This includes environmental and electrical parameters such as ambient temperature, relative humidity, and insulator surface leakage current. Next, the collected data are processed and analyzed using a smart algorithm that integrates multiple variables. This algorithm evaluates the pollution status and identifies patterns that indicate increasing flashover risk. Finally, the system outputs a real-time risk assessment to grid operation personnel, enabling timely preventive action. The entire system has been engineered with a focus on practical applicability in field conditions.

To support these functions, a distributed data acquisition architecture was adopted. This architecture comprises three primary components: (1) site monitoring terminals, responsible for real-time data collection; (2) a data processing unit, which performs preliminary data filtering and formatting; and (3) a backend analysis system, where advanced algorithms conduct comprehensive evaluation and flashover risk prediction.

The system achieves four key functional objectives: on-site data acquisition, wireless data transmission via 3G communication, real-time data analysis, and pollution risk evaluation. These functionalities work together to ensure accurate monitoring and timely warning of contamination flashover risks. An overview of the system structure and information flow is illustrated in Figures 11.1 and 11.2.

The flashover prewarning system is designed with several key features that enable continuous monitoring, risk assessment, and proactive management of pollution-related flashover hazards in power grids.

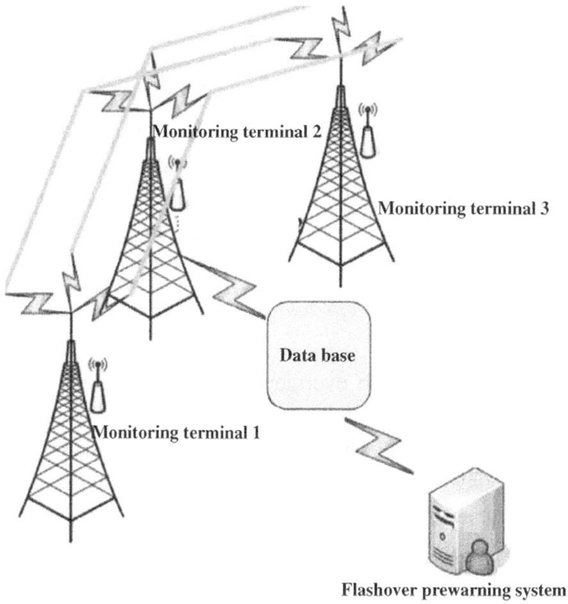

Figure 11.1 System setup. *Source*: Zhao et al. (2015)/with permission of IEEE.

Figure 11.2 Sensor implemented on the tower. *Source*: Zhao et al. (2015)/with permission of IEEE.

11.2.1 Digitized Maps of Polluted Areas in the Power Grid

A network of monitoring facilities has been deployed in regions identified as being prone to contamination. The system provides operators with access to digitized maps that display the geographical distribution of these monitoring sites. Through an intuitive user interface, real-time pollution data and contamination levels for each location are visualized. This spatial insight enables rapid identification of high-risk zones and supports informed decision-making for maintenance planning and operational control.

11.2.2 Pollution Flashover Warning

The system offers a comprehensive, real-time overview of each monitored site. For every location, the interface presents essential metadata including site name, geographical coordinates, voltage level, and the specific model of insulators installed. It also displays critical meteorological parameters such as ambient temperature, relative humidity, and rainfall. Leveraging this environmental data, the system calculates key insulation-related parameters, including leakage current, ESDD, pollution flashover voltage, and overall contamination severity. This integration of environmental and electrical metrics enables dynamic prediction of flashover risk, providing timely warnings to grid operators and supporting proactive measures to prevent insulation failures.

11.2.3 Weather Forecasting

Weather forecasting unit was integrated into the system, which would give warning information before weather as fog, rain and snow comes. Taking the weather forecast result and contamination state after a pollution accumulation period into consideration, this system could estimate the variation of insulation state in the next few days, and further achieve the purpose of pollution flashover prewarning.

11.3 System Application

From 2010 to 2014, multiple pollution flashover accidents occurred in Jiangsu Province, China. In response, over 180 leakage current monitoring devices were deployed across the Jiangsu power grid, particularly in regions identified as pollution-prone. The typical pollution-affected areas in Jiangsu can be categorized into three main types:

A. **Arid Regions in Northern Jiangsu**
   Xuzhou, a traditional mining and industrial city in northern Jiangsu, is home to numerous pollution sources, including coal mines, power plants, and cement

factories. As one of the most severely polluted areas in the province, Xuzhou has experienced frequent flashover incidents in the past.

B. **Coastal Power Transmission Corridor near Lianyungang**
This region hosts many salt chemical factories and is located near the coastline. The transmission line serving the Tianwan Nuclear Power Plant—specifically the Tiandu Line—is situated close to the sea, notably around towers 13# to 18#. A pilot monitoring system has been implemented in this area to provide real-time data on insulator contamination levels.

C. **Industrial Belt Along the Yangtze River in Central Jiangsu**
The cities of Nanjing, Yangzhou, and Zhenjiang have become hubs for heavy industry, including power generation, steel manufacturing, and chemical production. These industries contribute significantly to local pollution, making the region a key area of concern for flashover risk management.

The designed pollution monitoring system based on leakage current can effectively monitor changes in leakage current, while reflecting changes in pollution level and weather environment. Tower 276# on the Tiandu transmission line—serving as an outgoing tower for the Tianwan Nuclear Power Plant—was identified as a heavily polluted structure. To monitor the real-time pollution status of the line, a measurement system was installed on the B-phase insulator of the tower. This system collected environmental and electrical data, enabling detailed analysis of pollution behavior under changing weather conditions.

On November 4th, 2012, rainfall was recorded at Tower 276#. During this period, the ambient temperature ranged from 20 to 15 °C. Monitoring data—including temperature, relative humidity, and insulator leakage current—were analyzed to understand the impact of environmental factors on contamination activity. As shown in Figure 11.3, leakage current at midnight on November 3rd remained relatively stable at around 2 mA. Beginning at approximately 15:00, a notable drop in temperature was observed, accompanied by a sharp rise in relative humidity. These changes caused the pollution layer on the insulator surface to become moist, which in turn led to a rapid increase in leakage current.

Rainfall began at approximately 23:00, further contributing to the wetting of the pollution layer. The maximum leakage current, recorded at 8:10 on November 4th, peaked at 16.7 mA. Following this, a reduction in leakage current was observed, which is attributed to the thermal effect caused by the current itself. This heating effect produced localized dry areas on the insulator surface, temporarily reducing conductivity.

Throughout the period between 18:00 on November 3 and 8:10 on November 4, multiple drying and wetting cycles occurred on the insulator surface. These transitions directly influenced the leakage current behavior.

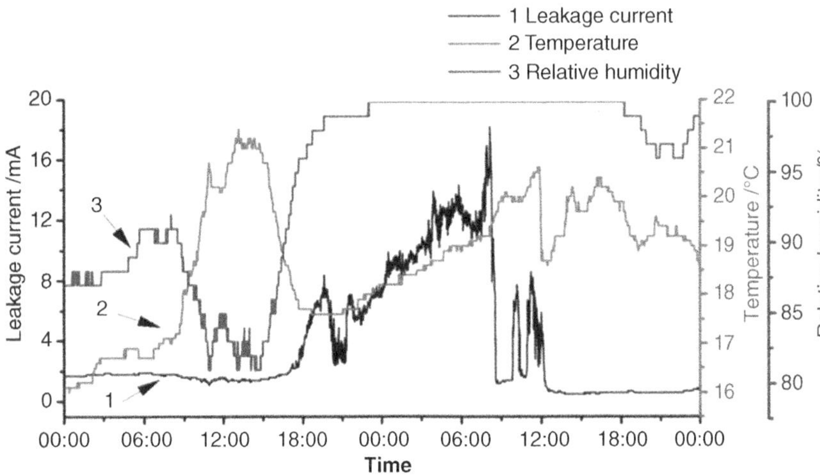

Figure 11.3 Monitoring data. (November 3 and 4, 2012). *Source*: Zhao et al. (2015)/with permission of IEEE.

After 8:10, the leakage current began to decrease significantly as the continued rainfall helped wash away surface contaminants, effectively reducing the pollution severity on the insulator.

This case illustrates the dynamic interaction between environmental conditions and insulator contamination behavior. It also highlights the effectiveness of real-time monitoring systems in capturing early warning signals and supporting timely operational decision-making.

11.4 Algorithms for Real-time Contamination Assessment

The real-time assessment of insulator contamination based on leakage current is a complex process, as the leakage current of glass or porcelain insulators is influenced not only by the degree of surface pollution but also by prevailing weather conditions. In particular, the moisture content of the pollution layer plays a critical role in determining leakage current behavior.

Previous studies have shown that the moisture characteristics of the pollution layer differ significantly between atmospheric conditions where the relative humidity is below 100% and when it is exactly 100%. Consequently, the correlation between ESDD—measured in mg/cm^2—and leakage current varies under these two moisture regimes.

Based on the solubility of the salts within the contamination layer, early warning methods that utilize leakage current for evaluating surface pollution on insulators can be classified into two main categories:

1. Saturated Moisture Condition
   In this scenario, the pollution layer on the insulator is fully saturated with moisture, and all soluble salts are completely dissolved. This condition is typically observed during or after heavy fog, dew, or rainfall. Under such circumstances, the leakage current is more directly and linearly related to the ESDD, making contamination assessment relatively straightforward.
2. Unsaturated Moisture Condition
   In contrast, when the pollution layer is exposed to humidity levels below saturation, only a portion of the soluble salts are dissolved. This results in a more complex and nonlinear relationship between leakage current and ESDD. In such cases, both the hygroscopic behavior of the contaminants and the environmental parameters—such as temperature and humidity—must be considered in the algorithm to obtain an accurate contamination assessment.

Understanding these two distinct conditions is essential for developing reliable early warning algorithms. A robust real-time contamination assessment system must account for the dynamic interaction between weather conditions and pollution behavior, particularly in transitional states between dry and fully saturated surfaces.

11.4.1 At Saturated Moisture

When the atmospheric humidity reaches 100%, the insulator moisture is mainly caused by the collision of water molecules, which leads to the moisture rate being much higher than the evaporation rate. In this case, the salt on the pollution layers surface is fully dissolved, the insulator pollution degree, ESDD is directly related to the magnitude of the leakage current, that is,

$$\mathrm{ESDD} = k_c \times f' \times \frac{I_{100\%}}{U} \tag{11.1}$$

where $I_{100\%}$ is the leakage current amplitude under saturated moisture; k_c is the correction coefficient of the contamination component and for the periphery of typical pollution sources, the coefficient does not change with the degree of pollution. f' is the corrected insulator shape factor, which considers the local arc effect on the leakage current distribution.

11.4.2 At Unsaturated Moisture

As discussed in Chapter 5, the wetting behavior of polluted insulators under unsaturated moisture conditions (relative humidity less than 100%) is significantly influenced by the site-specific pollution severity. While the accumulation of atmospheric contamination on insulator surfaces is a gradual process, the leakage current reacts quickly to short-term variations in ambient temperature and humidity. This fundamental difference in time scales—between the slow evolution of true pollution severity and the instantaneous electrical response to the environment parameters—provides a practical basis for distinguishing between the actual insulator pollution degree by leakage current trends.

The change in water content within the contamination layer follows an exponential approach function, which governs the transient behavior of leakage current. Specifically, the time required for the leakage current to rise from zero to its equilibrium value typically falls within the range of three to five time constants (3–5 τ). Considering the high-voltage stress applied to insulators in operation, the influence of this transient can often be neglected. For instance, based on the unified specific creepage distances of 22.0 to 53.7 mm/kV, a voltage of approximately 6.0 to 14.5 kV may be applied to a single cap-and-pin insulator with a leakage distance of 320 mm. Given that the time constant τ is typically on the order of a few seconds, the transition period is extremely short relative to operational timescales. Therefore, under steady environmental conditions, the measured leakage current on an online operating insulator can be reasonably assumed to represent the equilibrium value corresponding to the current atmospheric relative humidity (RH). This allows for the establishment of a functional relationship between the real-time leakage current $I(\mathrm{RH})$, the environmental humidity RH, and the ESDD of the insulator, expressed as:

$$I(\mathrm{RH}) = p \times (\mathrm{RH} - \mathrm{RH}_0) \tag{11.2}$$

where p and RH_0 are the fitted coefficients.

The parameter p is the slope between leakage current and humidity, which could also be calculated by,

$$p = \frac{\Delta H \times S \times f \times k_{\mathrm{RH}}}{U} \times \mathrm{ESDD} \tag{11.3}$$

As the slope p is mainly affected by the pollution degree of the insulator, it could be used to monitor the insulator. The hygroscopicity of different pollution components might be different, which may result in different $k_{\mathrm{RH\text{-}values}}$ under the same pollution degree. Fortunately, in the specific area with the same typical pollution source, the proportion of insulator contamination component does not change with time.

Because leakage current responds rapidly to environmental fluctuations while pollution builds up slowly, it is possible to assess the contamination level by analyzing the dynamic behavior of leakage current over a short period. By collecting a large volume of leakage current and environmental data over several hours or a single day, a characteristic parameter—represented by the slope p in Eq. (11.3)—can be extracted. This slope, which captures the correlation between leakage current variation and RH, is found to be proportional to the pollution degree.

Once the slope p is fitted using real-time measurements, it can be used to estimate the insulator's ESDD or insulator pollution flashover voltages. This method enables practical and responsive assessment of insulator pollution without the need for direct sampling or invasive measurements.

The overall process of contamination assessment under unsaturated conditions is illustrated in Figure 11.4, which outlines the steps of data acquisition, real-time correlation analysis, and contamination level estimation.

As illustrated in Figure 11.4, the procedure for assessing insulator contamination under unsaturated moisture conditions involves the following steps:

Real-Time Data Acquisition: Measure the leakage current and ambient environmental parameters—specifically, temperature and relative humidity—using online monitoring equipment installed on the insulator.

Temperature Correction: In accordance with the IEC 60815-1 standard, correct the measured leakage current to its equivalent value at 20 °C, ensuring consistency across varying thermal conditions.

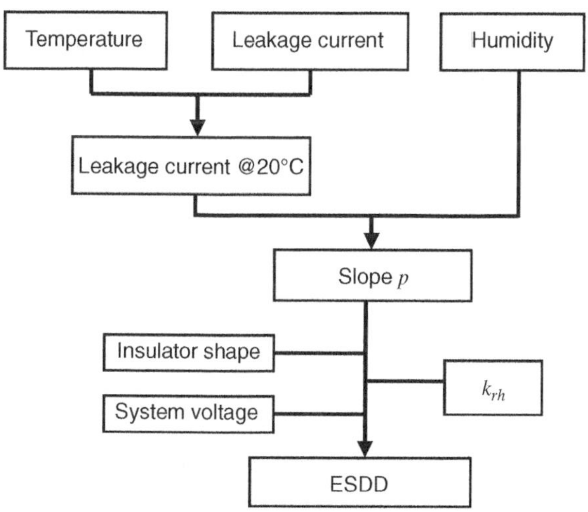

Figure 11.4 Process of insulator contamination assessment. *Source*: Cao et al. (2022)/ with permission of IEEE.

Data Fitting: Assuming the pollution level remains stable over a short time frame, apply curve fitting to hundreds of leakage current and RH data points collected during that period. Determine the fitting parameter p, which minimizes the fitting error. This parameter reflects the insulator's pollution-dependent wetting characteristics.

ESDD Estimation: Once the parameter p is obtained, use Eq. (11.3) to correct the applied voltage and current values. This enables the estimation of the insulator's ESDD, representing the actual contamination severity.

11.4.3 With Temperature Difference

Under certain meteorological conditions—particularly during clear, windless nights or when wind speeds are below 0.5 m/s—radiative cooling can cause the surface temperature of insulators to fall below the surrounding air temperature. This phenomenon can significantly accelerate the wetting process on insulator surfaces. Despite its practical relevance, the influence of temperature difference on insulator wetting has received limited attention in past studies. Some researchers have suggested that when a temperature gradient exists between the insulator and the ambient air, the moisture content of the contamination layer becomes closely related to the atmospheric dew point. Specifically, if the insulator surface temperature drops below the dew point, the pollution layer can rapidly reach saturated moisture conditions, even in the absence of direct precipitation or fog.

However, it is important to clarify that dew point temperature represents the threshold at which atmospheric water vapor, under constant pressure and humidity, transitions from an unsaturated to a saturated state. Therefore, while the dew point is a useful reference, it does not fully capture the complexity of the wetting process under radiative cooling conditions.

Experimental results reported in Chapter 5 further highlight this nuance. These tests showed that even when the insulator surface temperature was lower than the ambient air temperature but still above the dew point, the leakage current through the contaminated layer increased significantly. This indicates that the dew point temperature may be a sufficient but not necessary condition for insulator wetting in the presence of a temperature difference. These findings suggest that temperature difference effects should be considered in leakage current modeling and wetting prediction algorithms, especially for early-stage wetting detection in lightly polluted or marginally humid environments.

In the wetting process, the water absorption of surface contamination is related to the RH. The RH is a percentage, that indicates a present state of absolute humidity relative to a maximum humidity given the same temperature. When the insulator temperature is lower than that of the ambient, assuming the water content and pressure at the interface between air and contamination layers do not change suddenly and remain the same as those of the atmosphere, the interface's RH would

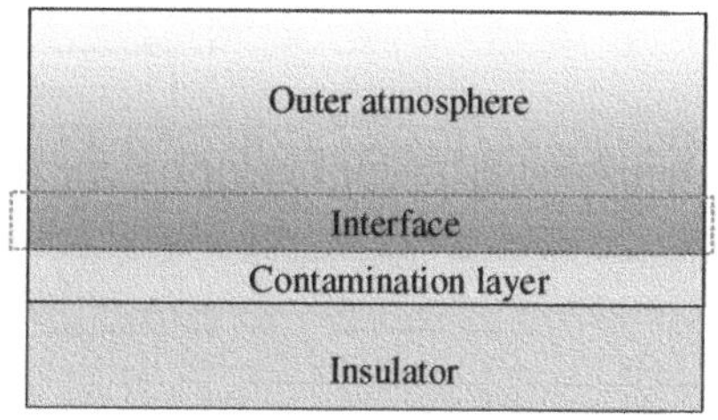

Figure 11.5 Contamination layer, interface, and atmosphere. *Source*: Cao et al. (2022)/with permission of IEEE.

change with the temperature, further changing the moisture rate. To briefly describe the wetting process with temperature difference, we divide the atmosphere above the contamination layer into two parts, namely the outer atmosphere and the interface between the atmosphere and pollution layers, as shown in Figure 11.5.

When there is a temperature difference between the insulator surface and the surrounding environment, the interface's humidity RH_{inter} is

$$RH_{inter} = \frac{\rho_w}{\rho_{w-max}} \times 100\% \tag{11.4}$$

where ρ_w and ρ_{w-max} represent absolute humidity and saturated absolute humidity respectively, $\rho_{w\text{-}max}$ could be calculated by,

$$\rho_{w-max} = \frac{M}{RT(temp)} p_s(temp) \tag{11.5}$$

where M is water molecular weight, 18 g/mol; R is gas constant, 0.0623665 mmHg·m^3/°C/mol; temp and T(temp) are the atmospheric temperature in Celsius and in Kelvin respectively; and p_s(temp) is the saturation vapor pressure. T(temp) and p_s(temp) could be calculated by:

$$\begin{cases} T(temp) = 273.15 + temp \\ p_s(temp) = (0.61078 \times 7.501)e^{\frac{17.2694temp}{238.3+temp}} \end{cases} \tag{11.6}$$

It should be noted that when the insulator surface temperature falls below the ambient dew point, the air near the surface becomes saturated with water vapor, and any excess moisture condenses to form fog droplets or dew. Under such conditions, the calculated interfacial RH, RH_{inter}, may exceed 100%. To account for this, we introduce the concept of "apparent humidity," which allows for humidity values greater than 100%. This apparent humidity represents the concentration of water molecules at the hypothetical insulator-air interface, and effectively reflects the degree of temperature difference between the surface and the surrounding atmosphere. A larger temperature difference results in a higher apparent humidity, indicating a faster moisture absorption rate by the contamination layer.

Once the apparent humidity of the pollution layer is determined, the method outlined in Section 11.4.1 can be employed to estimate the contamination level on

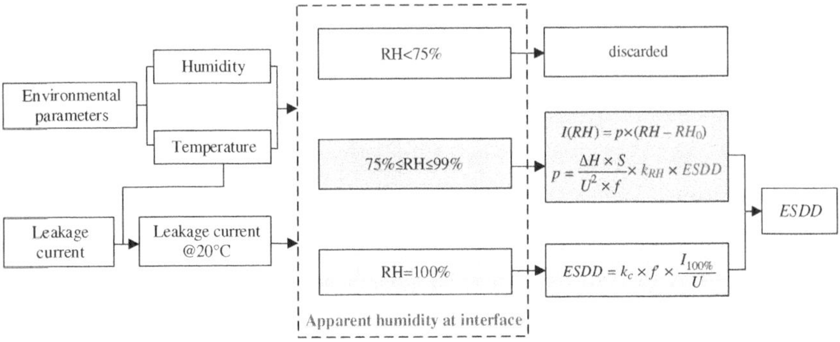

Figure 11.6 Online monitoring method of insulator contamination degree based on leakage current. *Source*: Cao et al. (2022)/with permission of IEEE.

the insulator surface. This approach offers a more accurate evaluation of pollution severity, particularly under conditions where temperature differences significantly influence wetting behavior. Subsequently, the degree of contamination can be assessed using Eq. (11.3) for RH levels between 75 and 100%, and Eq. (11.1) for conditions at 100% RH. The overall process for online monitoring of insulator contamination based on leakage current measurements is summarized in Figure 11.6.

11.5 Assessment Method of Pollution Degree for Hydrophobic Surface

In Chapter 10, we discussed the relationship between the wetting degree, pollution level, and flashover voltage on hydrophobic surfaces. For clarity, this relationship is redrawn in Figure 11.7. In the early stages of moisture exposure, the flashover voltage does not decrease significantly, so its impact on pollution flashover characteristics can be considered negligible. However, in the later stages—when a continuous water film form—the flashover voltage of hydrophobic surfaces becomes nearly indistinguishable from that of hydrophilic ones. Therefore, online monitoring of flashover voltage on hydrophobic surfaces is most critical during the intermediate stage of the wetting process.

As the insulator surface becomes progressively wetted, the distribution of wet and dry regions changes dynamically. With increasing moisture, the proportion of the wet area on the surface clearly increases. The remaining dry zones then experience intensified electric fields, which can lead to localized partial discharges. These discharges can be effectively detected by analyzing the ratio of fundamental and harmonic components in the leakage current signal. A physical model illustrating the changes in the polluted surface before and after localized discharge is shown in Figure 11.8.

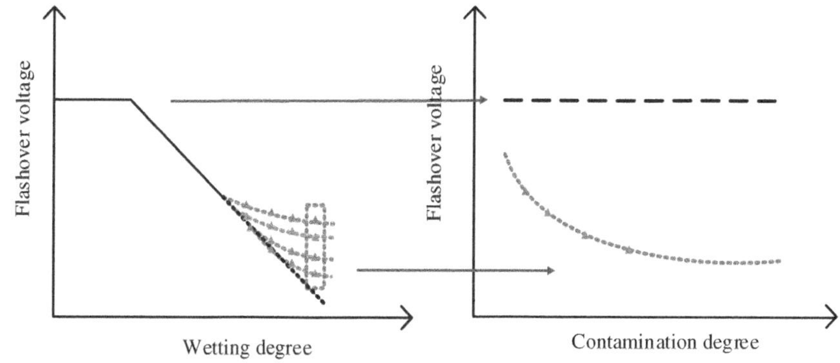

Figure 11.7 The trend diagram for relationship between flashover voltage contamination degree and wetting degree.

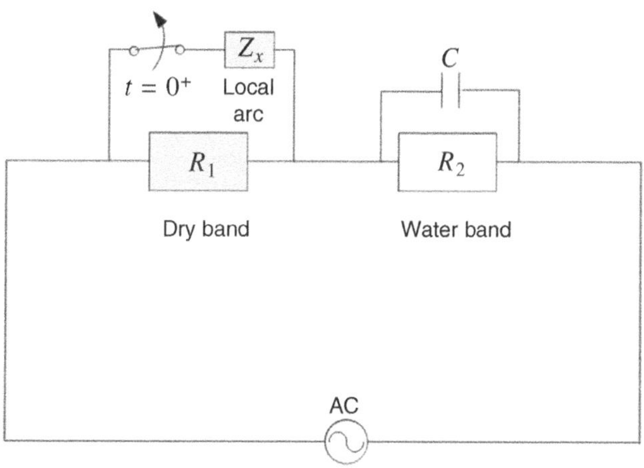

Figure 11.8 Analytical model of local discharge in polluted layers.

When the Joule heating effect of leakage current in a localized polluted area exceeds its moisture absorption capability, a local dry zone forms. In the equivalent circuit model, R_1 represents the resistance of the local dry zone, R_2 denotes the resistance of the remaining moist contamination layer, and C corresponds to the stray capacitance of the test specimen as well as the capacitance between water droplets. Since the stray capacitance is relatively small and presents a high impedance at power frequency, it can be treated as an open circuit during steady-state analysis.

The model assumes that R_1 and R_2 change only with the moisture content of the contamination layer. The transient discharge process has minimal impact on the state of the contamination layer, and the changes in R_1 and R_2 before and after discharge can be neglected.

At $t = 0^-$, it is assumed that a discharge spark bridges the dry band—equivalent to the switch being closed in Figure 11.8—producing a discharge pulse in the circuit. Furthermore, the combined impedance of Zx in parallel with R_1 is considered negligible compared to R_2. At $t = 0+$, the switch opens, and the discharge pulse begins to decay. Based on circuit theory, the behavior of this discharge process can be analyzed accordingly.

At $t = 0+$, the circuit current is given by:

$$\dot{I}_{0^+} = \dot{I}_{0^-} \approx \frac{\dot{U}}{R_2} \tag{11.7}$$

Once the discharge has fully decayed, the circuit current becomes:

$$\dot{I}_{\infty} = \frac{\dot{U}}{(R_1 + R_2)} \tag{11.8}$$

Therefore, the current in the circuit during the decay of the discharge pulse, from its peak value to a steady state, is expressed as:

$$\dot{I} = \dot{I}_{\infty} + \left(\dot{I}_{0^+} - \dot{I}_{\infty}\right)e^{-\frac{t}{\tau}} = \frac{\dot{U}}{R_1 + R_2} + \frac{R_1\dot{U}}{(R_1 + R_2)R_2}e^{-\frac{t}{\tau}} \tag{11.9}$$

In this expression, τ is the decay constant, which depends on the type of discharge and the characteristics of the circuit.

According to Eq. (11.9), after a discharge pulse occurs in the circuit, the fundamental (power frequency) component of the current consists of two parts:

- One part determined by the resistance of the pollution layer
- Another part introduced by the discharge pulse itself

By performing a Fourier transform on the exponential decay function (of the form eat), the fundamental frequency component within the discharge pulse can be extracted.

$$\begin{aligned} F(\omega) &= \int_{-\infty}^{\infty} e^{at} e^{-\mathrm{j}\omega t} \mathrm{d}t = \int_{0}^{\infty} e^{(a-\mathrm{j}\omega)t} \mathrm{d}t \\ &= \left.\frac{e^{(a-\mathrm{j}\omega)t}}{a - \mathrm{j}\omega}\right|_0^{\infty} = \frac{-1}{a - \mathrm{j}\omega} = \frac{1}{\sqrt{a^2 + \omega^2}} e^{\mathrm{j}\arctan\frac{\omega}{a}} (a > 0) \end{aligned} \tag{11.10}$$

$$\varphi(\omega) = \arctan\frac{\omega}{a} \tag{11.11}$$

$$|F(\mathrm{j}\omega)| = \frac{1}{\sqrt{a^2 + \omega^2}} \tag{11.12}$$

In Eq. (11.9), the fundamental current in the circuit is given by:

$$\dot{I}_1 = \frac{\dot{U}}{R_1 + R_2} + \frac{R_1\dot{U}}{(R_1 + R_2)R_2} \frac{1}{\sqrt{a^2 + \omega^2}} \approx \frac{\dot{U}}{R_1 + R_2} \tag{11.13}$$

After a local discharge occurs within the pollution layer, the maximum total circuit current appears at $t = 0^+$, the moment the discharge spark fully bridges the local dry zone. At this instant, the current amplitude is determined by the equivalent resistance of the remaining moist pollution layer (R_2).

Due to the nonlinear nature of the discharge, a significant number of harmonic components are generated during the decay of the discharge pulse. When this decay process ends, the total circuit current stabilizes and equals the fundamental current, determined by the total resistance of the pollution layer ($R_1 + R_2$).

When localized discharge occurs in the pollution layer of the specimen, the ratio of the circuit's fundamental current to the total current is given by:

$$\frac{\dot{I}_1}{\dot{I}} = \frac{1}{1 + \frac{R_1}{R_2} e^{-\frac{t}{\tau}}} \tag{11.14}$$

The maximum ratio of fundamental current to total current before and after discharge is expressed as:

$$\frac{I_{1h}}{I_h} = \left[\frac{1}{1 + \frac{R_1}{R_2} e^{-\frac{t}{\tau}}} \right]_{t=0^+} = \frac{R_2}{R_1 + R_2} \tag{11.15}$$

According to the model shown in Figure 11.8 and Eq. (11.15), this ratio essentially reflects the intensity of surface discharges on the polluted layer, which in turn depends on the voltage distribution along the insulator surface. Based on Figure 11.8, if we assume the equivalent resistance of the discharge region (R_1) remains constant, the remaining pollution resistance (R_2) will vary significantly depending on the surface hydrophobicity.

- The better the hydrophobicity, the larger the value of R_2, resulting in a smaller voltage drop across R_1, leading to weaker discharge activity. As a result, the fundamental current becomes closer in magnitude to the total current, and their trend lines tend to overlap.
- Conversely, for surfaces with lower hydrophobicity, the discharge is stronger, and a larger deviation between the fundamental current and total current is observed.

Therefore, under wetting conditions, leakage current and its fundamental component exhibit distinct trend curves for specimens with different hydrophobic characteristics. Since the ratio of the fundamental-to-total leakage current reflects discharge intensity—which is closely tied to the pollution state—we define a parameter:

K_h – the Ratio of Peak Fundamental Component to Peak Total Current within a Given Time Interval

Figure 11.9 shows a set of measured curves with typical characteristics. It is important to note that the dashed box area in the figure corresponds to a K_h value that lacks significance. This is because, in the early stages of wetting, the pollution layer is still relatively dry, and the leakage current remains extremely low—often below the detection threshold of the monitoring system. During this phase, Gaussian white noise in the signal is comparable in magnitude to the actual leakage current, and contains a large number of harmonic components, which leads to artificially low K_h values.

The initial increase in K_h during early wetting is driven by two main factors:

1. As the wetting process progresses, the leakage current amplitude gradually enters the measurable range of the monitoring system, reducing the influence of background noise.
2. The pollution layer becomes more uniformly wetted, which promotes a more even electric field distribution and suppresses surface discharge activity.

The parameter K_h corresponds to surface discharge activity on the pollution layer. Since discharges are intermittent in nature, K_h values exhibit frequent fluctuations, making the raw K_h trend curve unsuitable for direct use in flashover voltage monitoring. Therefore, further processing is required.

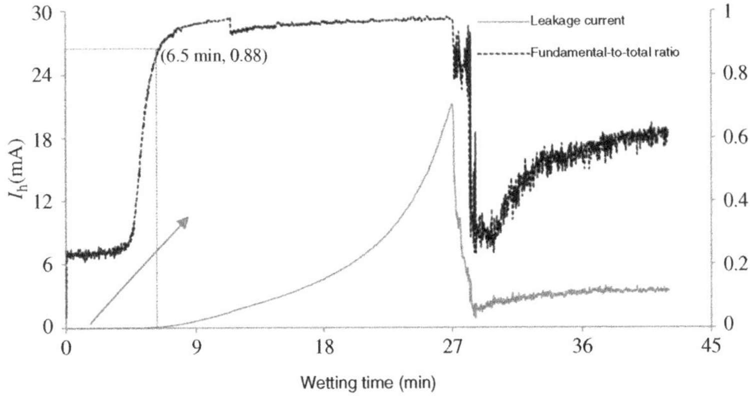

Figure 11.9 Trend of leakage current and fundamental-to-total wave ratio during moisture absorption on a hydrophobic surface (SDD: 0.1 mg/cm^2, NSDD: 1.0 mg/cm^2, hydrophobicity transfer time: 52 h).

In the subsequent analysis, we statistically evaluate K_h over a defined time interval T. Specifically, we compute the average of the 10 lowest K_h values within each period T, and denote this average as K_{h10}. This value, referred to as the Characteristic Fundamental-to-Total Ratio (K_{h10}), is used to represent the pollution condition of the insulator surface during the given time interval T.

For the data shown in Figure 11.9, all data points where the leakage current amplitude (I_h) is less than 0.1 mA are considered invalid and excluded from the analysis. By replacing the fluctuating K_h values with the more stable K_{h10}, a refined K_{h10} characteristic curve is obtained, as shown in Figure 11.10.

The parameter Kh_{10} characterizes discharge intensity during the wetting process and is closely related to both contamination severity and moisture absorption. Since an insulator's pollution flashover voltage is directly affected by these two factors, it is reasonable to investigate whether Kh_{10} can serve as a practical indicator for flashover voltage prediction.

To explore this relationship, ten kaolin-contaminated specimens with hydrophobicity transfer times ranging from one to two hours were tested. The pollution flashover voltage of each specimen was evaluated using the step-up voltage method. For specimens with a NSDD of 1.0 mg/cm^2, test results under both horizontal and vertical mounting conditions were aggregated and replotted in Figure 11.11. The correlation between the normalized characteristic parameter (Kh_{10}) and the flashover voltage gradient remained strong.

As shown in Figure 11.11, the flashover voltages of silicone rubber specimens tend to converge when the pollution layer conditions are similar. The state of the pollution layer—determined by a combination of moisture content and contamination level—can thus be effectively represented by the characteristic parameter $K^* h_{10}$.

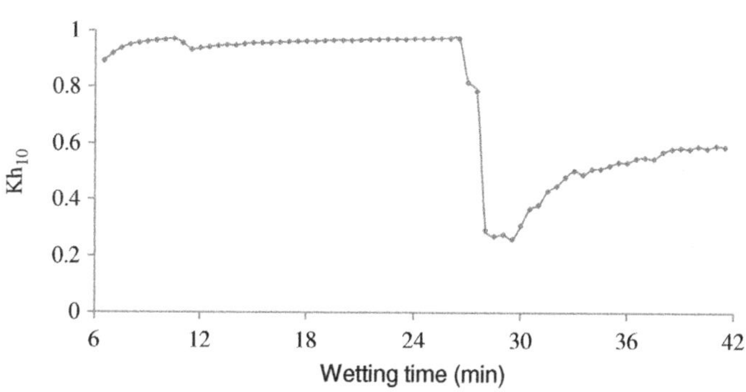

Figure 11.10 The full trend of k_{h10} over the entire wetting period is termed the k_{h10} characteristic curve.

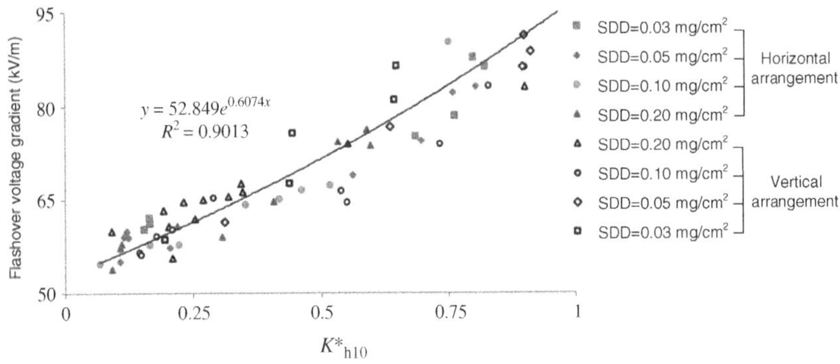

Figure 11.11 Summary of K^*_{h10} and flashover voltage gradient with different SDD (NSDD = 1.0 mg/cm^2).

11.6 Summary

The theoretical surface conductivity of the insulator under the saturated moisture condition is independent of measurement methods and it is only related to the amount of soluble salts dissolved in the contamination. However, it is found that the surface conductivity deduced from the leakage current is different from the theoretical surface conductivity.

The main cause is that, during the wetting process of the insulator in actual operation, a dry band and partial electric arc will be generated around the steel pin. Although the electric arc resistance is very small and even negligible, the path of the current on the residual contamination layer is changed, resulting in the uneven distribution of the leakage current on the insulator surface.

The investigation results have shown that, the surface conductivity converted from the leakage current is in proportion to the actual surface conductivity of the contaminated layer. However, for insulators with different profiles, the influence of partial electric arc can be different. The ratio of surface conductivity of the XWP-70 insulator under the saturated moisture condition to SDD is in a range of 750–850, while that of XP-70 insulator is roughly between 350–550.

For hydrophobic surfaces, the K_{h10} value reflects the intensity of surface discharge, which is fundamentally linked to the moisture level and contamination severity of the pollution layer. Studies have shown that specimens with different contamination levels and varying initial hydrophobic properties tend to exhibit similar flashover voltages as long as their K_{h10} values converge under fully wetted conditions. Since K_{h10} represents the real-time state of the pollution layer after moisture absorption, it demonstrates a strong correlation with the specimen's flashover voltage.

Bibliography

Amarh, F., Karady, G.G., and Sundararajan, R. Linear stochastic analysis of polluted insulator leakage current, IEEE Trans. Power Del., 2002, 17 (4), 1063–1069.

Chrzan, K. Conductivity of aqueous dust solutions, IEEE Trans. Electr. Insul, 1987, 22, 241–244.

Dong, H., Xu, W., Cao, B. et al. Effect of soluble salt on conductivity of partial surface, Annual Report IEEE Conference on Electrical Insulation and Dielectric Phenomena (CEIDP), 2015, pp. 185–188.

Farzaneh, M. and Chisholm, W. Insulators for Icing and Polluted Environments, Wiley, 2009.

Fierro-Chavez, J.L., Ramirez-Vazquez, Z., and Montoya-Tena, G. On-line leakage current monitoring of 400 kV insulator strings in polluted areas, IEE Proc. Gener. Transm. Distrib., 1996, 143 (6), 560–564.

GB/T 4585.2-1991, Artificial Pollution Tests on High-Voltage Insulators to be Used on AC Systems Solid-Layer Method, 1991. [in Chinese].

Guan, Z.C. External Insulation of Insulator and Power Transmission Equipment, Beijing, China: Tsinghua University Press, 2006. [in Chinese].

Holtzhausen, J.P. AC pollution flashover models: accuracy in predicting insulator flashover, Proceedings of the 10th International Symposium on High Voltage Engineering, 1997.

Holtzhausen, J.P. and Vosloo, W.L. The pollution flashover of AC energized post type insulators: the role of shortening of the arc, IEEE Trans. Dielectr. Electr. Insul., 2001, 8 (2), 191–194.

IEC 60507-2013, Artificial Pollution Tests on High Voltage Insulators to be Used on AC systems, 2013.

IEC/TS 60815-1, Selection and Dimensioning of High-voltage Insulators Intended for Use in Polluted Conditions-Part 1: Definitions, Information and General Principles, 2008.

IEEE Std. 4-2013, IEEE Standard Techniques for High-Voltage Testing, IEEE Publication, 2013.

Jiang, X., Yan, S., Sun, C. et al. Evaluating the safety condition of porcelain insulators by the time and frequency characteristics of LC based on artificial pollution tests, IEEE Trans. Dielectr. Electr. Insul., 2010, 17 (2), 481–489.

Li, J., Guan, Z., Wang, L. et al. An experimental study of AC arc propagation over a contaminated surface, IEEE Trans. Dielectr. Electr. Insul., 2012, 19 (4), 1360–1368.

Li, J., Sun, C., Sima, W. et al. Contamination level prediction of insulators based on the characteristics of leakage current, IEEE Trans. Power Del., 2010, 25 (1), 417–424.

Obenaus, F. Fremdschichtueberschlag und Kriechweglaenge, Disch. Elektrotechnik, 1958, 12 (4), 135–136.

Rizk, F.A.M. A criterion for AC Flashover of Polluted Insulators, IEEE Power Engineering Society Winter Power Meeting, New York, NY, USA, 1971.

Su, Z.Y. and Li, Q.F. Historical review and summary on measures against pollution flashover occurred in power grids in China, Power Syst. Technol., 2010, 34 (12), 124–130. [in Chinese].

Takasu, K., Shindo, T., and Arai, N. Natural contamination test of insulators with DC voltage energization at inland areas, IEEE Trans. Power Del., 1988, 3 (4), 1847–1853.

Wang, L., Cao, B., Mei, H. et al. Effects of natural contamination components on the surface conductivity under saturated moisture, IEEE Trans. Dielectr. Electr. Insul., 2017, 24 (5), 2945–2951.

Zhang, Z., Zhang, D., Jiang, X. et al. Study on natural contamination performance of typical types of insulators, IEEE Trans. Dielectr. Electr. Insul., 2014, 21 (4), 1901–1909.

Zhang, Z.J., Jiang, X.L., Huang, H.Z. et al. Study on the wetting process and its influencing factors of pollution deposited on different insulators based on leakage current, IEEE Trans. Power Del., 2013, 28 (2), 678–685.

Zhao, C.L., Mei, H.W., Zhu, M.X. et al. Development of contamination flashover pre-warning system and analysis of operating experience, IEEE Trans. Electr. Insul., 2015, 22 (4), 2234–2241.

12

Insulator Contamination Monitoring at Saturated Moisture

12.1 Introduction

Building on the discussion of contamination monitoring under unsaturated humidity in the previous chapter, this chapter focuses on methods applicable under saturated (100%) humidity conditions. The maximum leakage current observed on an insulator is strongly correlated with the degree of surface contamination. Therefore, if this maximum leakage current—measured under saturated moisture conditions—can be converted into surface conductivity by incorporating the operating voltage and the geometry of the insulator, it becomes feasible to implement a practical online monitoring method for insulator contamination.

However, during the wetting process, the thermal effect of the leakage current causes the formation of dry bands on the insulator surface. These dry bands increase the surface resistance and can disrupt the accurate reflection of pollution severity based on leakage current alone. Nevertheless, since dry bands exhibit high resistance and are subjected to strong electric fields, partial electric arcs may form across them when the local electric field exceeds a critical threshold. These arcs, due to their relatively low arc resistivity, allow leakage current to flow and can, under certain conditions, reflect the conductive properties of the remaining polluted areas. However, this behavior requires further experimental verification to be fully validated.

In this study, two types of porcelain insulators and a triangular glass plate were selected as test specimens. Their leakage current behavior during the wetting process under operating voltage was recorded and analyzed. The surface conductivity of each sample was calculated based on the maximum leakage current, applied voltage, and specific insulator geometry.

The results indicated that, due to the presence of partial arcs, the conductivity values estimated from leakage current were generally lower than the actual surface conductivity of the contamination layer. Despite this underestimation, for certain insulator types, a relatively stable and predictable relationship between the measured and actual values was observed. These findings offer valuable reference data for the development of real-time online contamination monitoring methods, particularly under saturated moisture conditions.

12.2 Test Specimens, Facilities, and Procedure

The experimental setup was designed to evaluate the leakage current behavior of insulators under saturated moisture conditions, and primarily consisted of three components: a pollution test power supply, an artificial fog chamber, and a leakage current monitoring system.

The pollution test power supply had a capacity of 100 kVA, with a maximum output voltage of 50 kV and a rated output current of 2 A, providing stable and controllable high-voltage conditions for the insulator samples during testing.

The leakage current monitoring system was capable of measuring currents in the range of 0.5 mA to 5 A, with an error probability of less than 3%. The system operated at a sampling frequency of 5 kHz, which allowed for accurate and real-time tracking of rapid changes in leakage current during the wetting and arcing processes.

The artificial fog chamber was constructed using an insulating plexiglass cylinder with a height of 120 cm and a diameter of 84 cm. Micron-sized water mist was generated by an ultrasonic humidifier and introduced into the chamber to create a controlled saturated moisture environment for the contamination layer. This setup ensured uniform fog distribution and consistent test conditions throughout the experiment.

A schematic diagram of the complete test system, including the power supply, fog chamber, and monitoring instrumentation, is provided in Figure 12.1.

The experimental specimens selected for this study included a triangular glass plate and two types of porcelain insulators, namely XWP-70 and XP-70. The triangular glass plate was isosceles in shape, with a height of 45 cm and a base length of 30 cm. To facilitate voltage application, both the top and bottom surfaces of the specimens were treated with conductive paper, ensuring uniform electrical contact.

The shape, dimensions, and key physical parameters of the test samples are illustrated in Figure 12.2 and summarized in Table 12.1. The form factor of the test sample is defined as follows:

$$f = \int_0^L \frac{dx}{\pi D(x)} \tag{12.1}$$

where L is the creepage distance of insulators, and $D(x)$ refers to the diameter at the area where the creepage distance is x.

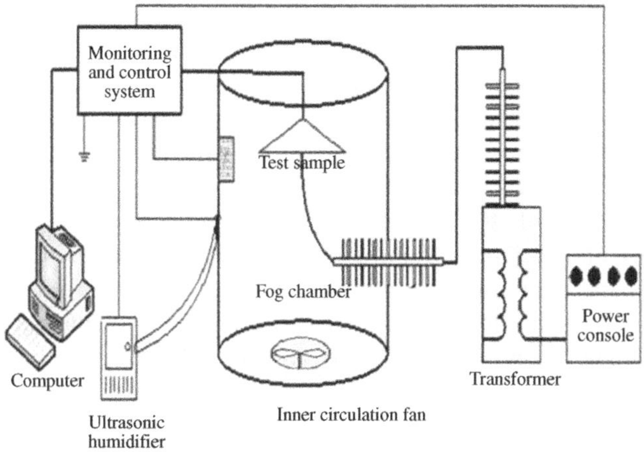

Figure 12.1 Schematic diagram of the experiment system. *Source*: Adapted from Cao et al. (2018).

Figure 12.2 Test samples: (a) XP-70; (b) XWP-70; (c) the triangle glass plate. *Source*: Cao et al. (2018)/with permission of IEEE.

Table 12.1 Parameters of tested samples.

Type	Creepage distance (mm)	Surface area (cm^2)	Form factor
Triangle glass plate	385	609	3.386
XWP-70	400	2023	1.104
XP-70	295	1591	0.736

Source: Cao et al. (2018)/with permission of IEEE.

The solid layer method was employed to create uniform contamination layers on the surface of the test specimens prior to the experiments. In accordance with standardized procedures, sodium chloride (NaCl) and kaolin were used as the soluble salt and non-soluble component, respectively. The contamination preparation followed the guidelines provided in IEC 60507, IEEE Std. 4, and GB/T 4585-2004. The preparation process was as follows:

First, quantitative amounts of NaCl and kaolin were precisely measured using a high-precision analytical balance. The dry materials were then mixed with an appropriate volume of distilled water to form a homogenous suspension. This mixture was applied evenly onto the test specimen surfaces using a small brush, ensuring full coverage and uniform thickness. The coated specimens were then left to dry completely under controlled conditions.

In this experiment, the Non-soluble Deposit Density was fixed at 1.0 mg/cm^2, while the Equivalent Salt Deposit Density (ESDD) was varied across five levels: 0.025, 0.05, 0.1, 0.15, and 0.2 mg/cm^2.

To simulate realistic operating conditions, the applied voltage for both the porcelain insulators and the triangular glass plate was set to 8 kV, which corresponds to the average voltage experienced by a single insulator in a transmission line. For the triangular glass plate, this voltage level also ensured the formation of partial arcs after saturation.

Given that transmission line insulators operate continuously under voltage, the test specimens were energized during the wetting process. After the contamination layers were prepared and fully dried, the insulators were placed in the artificial fog chamber. High voltage was applied to both ends of the specimens, and the ultrasonic humidifier was activated to increase the relative humidity within the chamber. The leakage current was monitored in real time throughout this process to capture the electrical response of the contaminated surface under saturated moisture conditions.

To determine the wetting duration required for the test samples, a preliminary experiment was conducted in which a polluted insulator was placed inside the artificial fog chamber. The surface conductivity trend during the wetting process was measured using an LCR electric bridge. To ensure accurate and stable measurements, the connection points at both ends of the insulator were coated with conductive adhesive, allowing full electrical contact between the measuring electrodes and the contamination layer.

A low measurement voltage of 2 V was applied during this process, which was insufficient to generate dry bands, ensuring that the measurement reflected the continuous wetting behavior of the pollution layer.

The results showed that the surface conductivity increased steadily and reached its maximum value after approximately 12 minutes. Beyond that point, a gradual decline in conductivity was observed, indicating the onset of contaminant wash-off or redistribution on the insulator surface. This trend is illustrated in Figure 12.3.

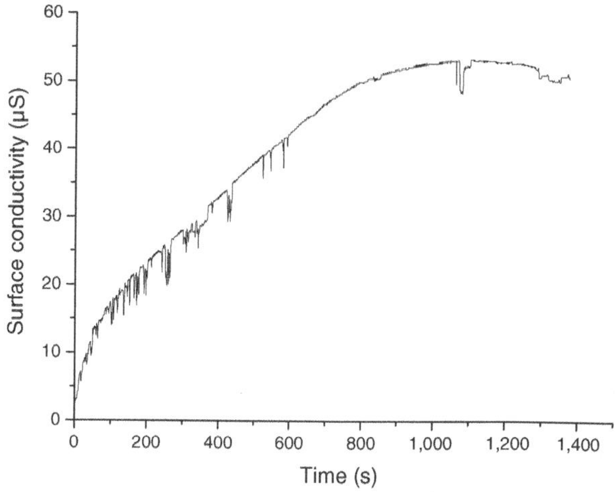

Figure 12.3 The tendency of surface conductivity during the wetting process (ESDD = 0.05 mg/cm^2). *Source*: Cao et al. (2018)/with permission of IEEE.

Based on this observation, the standard test duration for subsequent experiments was set to 15 minutes, ensuring that measurements captured the period of peak surface conductivity while minimizing the influence of pollutant loss.

12.3 Experimental Results

Some typical post-experiment conditions of the contamination layer on the test samples are illustrated in Figure 12.4, where the ESDD was set to 0.15 mg/cm^2. As shown in the figure, following a period of wetting under energized conditions, dry bands were observed forming near regions with the minimum equivalent radius—notably at the tip of the triangular glass plate and around the steel pin of the porcelain insulator.

These dry bands are the result of localized heating due to leakage current concentration, which leads to partial drying of the pollution layer despite the saturated moisture environment. Furthermore, after the contamination layer had completely dried, visible traces of electric arcs been found on the surface. These arc marks indicate that partial discharges occurred during the experiment, likely initiated at the dry band locations due to the intensified electric field and breakdown of the residual moisture path.

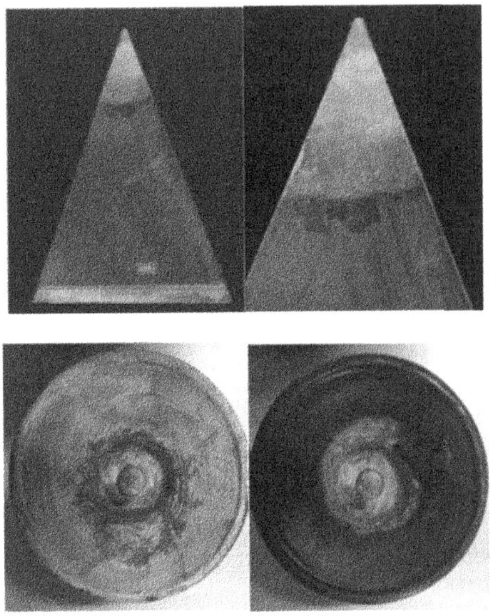

Figure 12.4 Test samples after the experiment. *Source*: Cao et al. (2018)/ with permission of IEEE.

12.3.1 Leakage Current Trend

During the wetting process, the maximum value of the leakage current measured within each second was recorded for analysis. The leakage current trends for different test samples under an ESDD of 0.1 mg/cm^2 are illustrated in Figure 12.5. As shown in the figure, numerous current pulses occurred throughout the wetting process. These pulses were primarily attributed to surface discharges and the formation of partial electric arcs.

The maximum leakage current values for each test sample under various levels of surface deposit density (SDD) were summarized and presented in Table 12.2.

From the experimental results, it was observed that for a given test sample, the maximum leakage current during the wetting process increased with higher SDD levels. However, when comparing different test samples under the same SDD, noticeable differences in maximum leakage current were recorded. This variation is likely due to differences in the geometry and surface profile of the samples, which affect the electric field distribution and moisture behavior.

To further evaluate the pollution severity of each insulator, the surface conductivity of the contamination layer was calculated. This calculation was based on the measured maximum leakage current, in combination with the applied voltage and the geometric form factor of each test specimen.

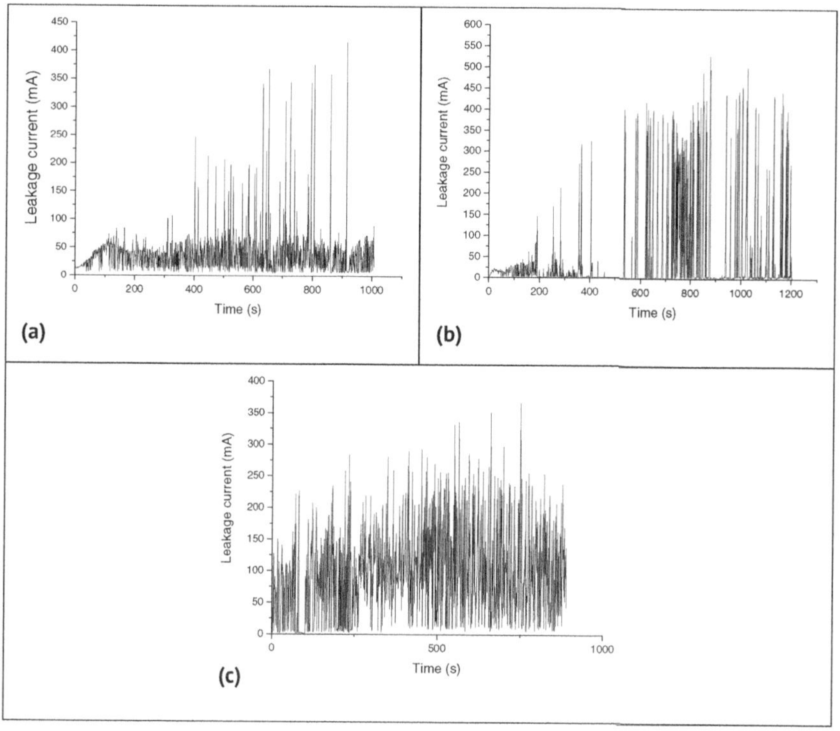

Figure 12.5 Leakage current value of each test sample (ESDD = 0.1 mg/cm^2): (a) XP-70; (b) XWP-70; (c) the triangle glass plate. *Source*: Cao et al. (2018)/with permission of IEEE.

Table 12.2 Maximum leakage current during the wetting process.

SDD (mg/cm^2)	Triangle glass plate (mA)	XWP-70 (mA)	XP-70 (mA)
0.025	76	153	120
0.05	160	291	236
0.10	367	529	416
0.15	565	925	906
0.20	650	1130	1131

Source: Cao et al. (2018)/with permission of IEEE.

12.3.2 Surface Conductivity

The maximum value of the leakage current measured during the wetting process can be converted to surface conductivity, and the surface conductivity of the sample with different SDD was calculated by Eq. (12.2)

$$K = \frac{f \times I}{U} \tag{12.2}$$

where K is the surface conductivity of insulators, f is the form factor of the test samples, I is the maximum leakage current of the test samples, and U is the voltage applied.

The surface conductivity of the test samples obtained from under different SDD is shown in Figure 12.6.

It was observed that there was an approximately linear relationship between the surface conductivity, as deduced from the maximum leakage current, and the SDD. However, for different test samples, the surface conductivity values corresponding to the same SDD were found to differ.

Theoretically, under saturated moisture conditions, the surface conductivity of a contamination layer should depend solely on the amount of soluble pollutants

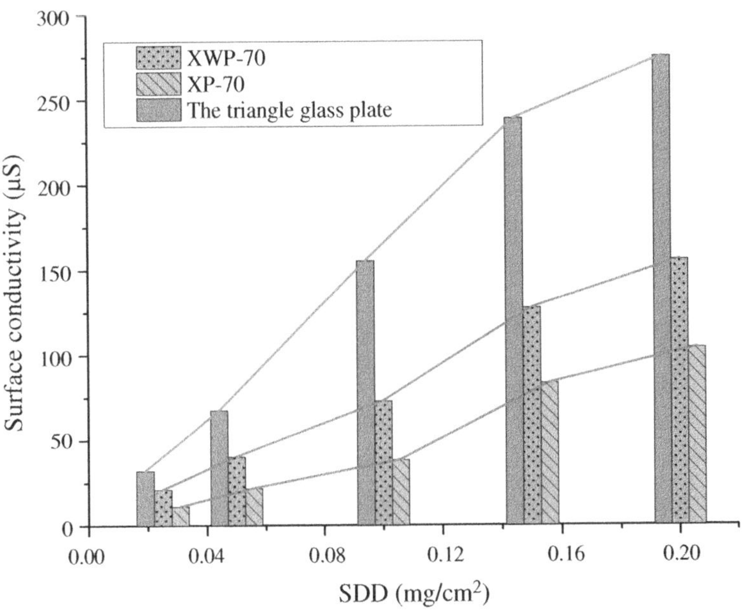

Figure 12.6 Relationship between the surface conductivity calculated using leakage current and SDD. © 2017 IEEE.

dissolved in the moisture film. In this ideal case, it should be independent of the insulator's shape or structural geometry.

The inconsistencies found in the experimental data suggest that additional factors influence the conductivity measurements in practice. The possible causes of these differences are analyzed in the following section.

12.4 Analyses and Discussion

12.4.1 Theoretical Relationship Between the Surface Conductivity and SDD

Theoretically, the surface conductivity of a contamination layer is determined primarily by two factors: the degree of wetting and the SDD. During the wetting process, as more soluble salts in the contamination dissolve, the ion concentration in the resulting surface solution increases. This leads to a corresponding increase in surface conductivity.

The relationship is straightforward: the greater the amount of dissolved salt, the higher the ionic content, and thus the higher the electrical conductivity of the surface layer. Based on this principle, the theoretical surface conductivity for different SDD levels can be calculated using the method illustrated in Figure 12.7.

Based on the conductivity of contaminated solutions with different concentrations of NaCl, the relationship between surface conductivity and water absorption of the contamination layer can be obtained, as is shown in Figure 12.8. The surface conductivity is saturated with the increase of water absorption.

The relationship between surface conductivity under the saturated moisture condition and SDD is shown in Eq. (12.3).

$$K = 987 \times \text{SDD} + 9.4 \tag{12.3}$$

where K is the surface conductivity of the contamination layer under saturated moisture. The surface conductivity is roughly 1000 times that of SDD.

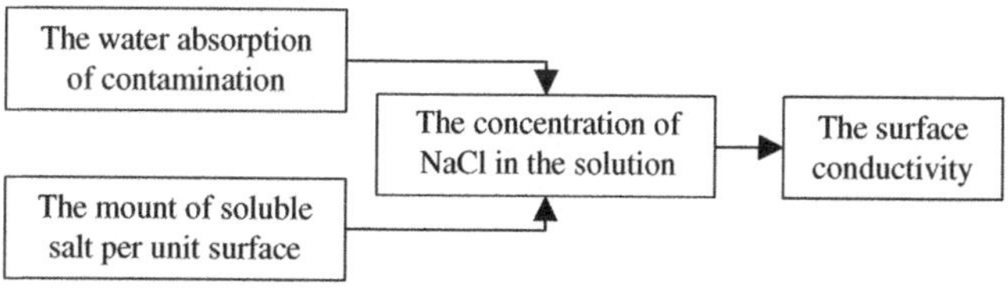

Figure 12.7 Flow chart of theoretical surface conductivity calculation.

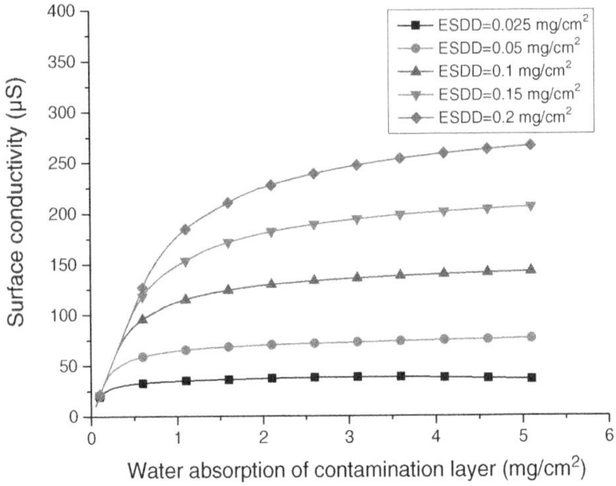

Figure 12.8 The relationship between surface conductivity and water absorption of contamination layer. *Source*: Cao et al. (2018)/with permission of IEEE.

12.4.2 Influence of Local Electric Arc

In this study, leakage current was used as the primary parameter for online monitoring, and the insulators remained energized throughout the measurement process. As a result, due to the thermal effect of the current flow, dry bands inevitably formed on the surface of the insulators during the wetting process. Under such conditions, it was evident that the leakage current could only reflect the true conductive state of the contamination layer—and thus the pollution severity—if partial electric arcs formed to bridge the dry areas. These arcs effectively restored the current path and allowed the measured leakage current to represent the conductivity of the residual wet contamination layer.

Figure 12.9 presents the ratio of surface conductivity to SDD under saturated moisture conditions for two types of insulators. For the XP-70 insulator, this ratio ranged approximately between 750 and 850, whereas for the XWP-70, the ratio was notably lower, ranging from 350 to 550.

Interestingly, both of these measured ratios deviate significantly from the theoretical expectation, in which the surface conductivity is predicted to be approximately 1000 times the SDD. This discrepancy suggests that practical factors—such as nonuniform wetting, dry band formation, and electric field distribution differences—may significantly influence the measured values and must be accounted for in any real-time contamination assessment model.

For evenly distributed contamination on an insulator surface, the leakage current path under saturated moisture conditions—in the absence of dry

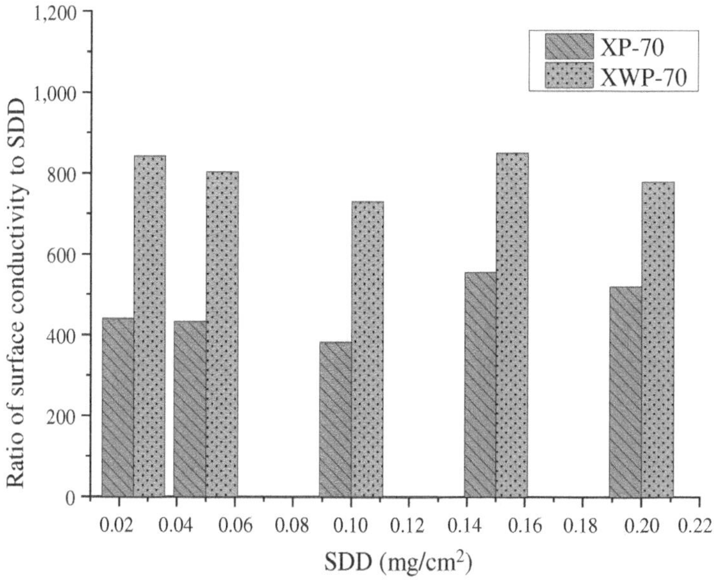

Figure 12.9 The ratio of surface conductivity to ESDD.

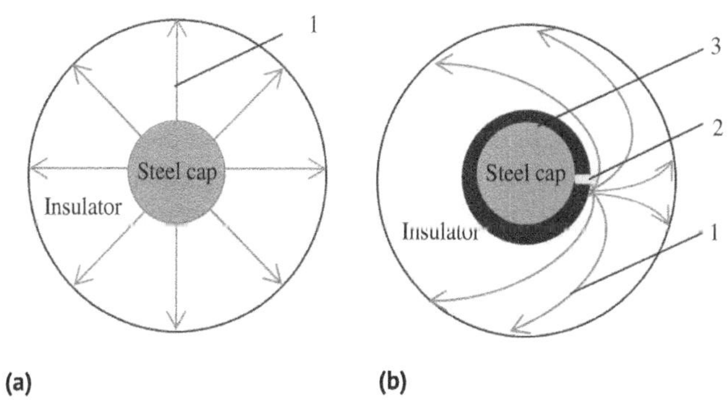

Figure 12.10 Flow path of surface leakage current on the insulator's surface: (a) without dry band; (b) with dry band and partial electric arc. 1: Current path; 2: partial electric arc; 3: dry band. *Source*: Cao et al. (2018)/with permission of IEEE.

bands—is shown in Figure 12.10a. However, when dry bands form and are bridged by partial electric arcs, the current path is significantly altered, as illustrated in Figure 12.10b. This change in current flow is functionally equivalent to modifying the geometry of the insulator, as it alters the distribution of the electric field and the effective current path along the surface.

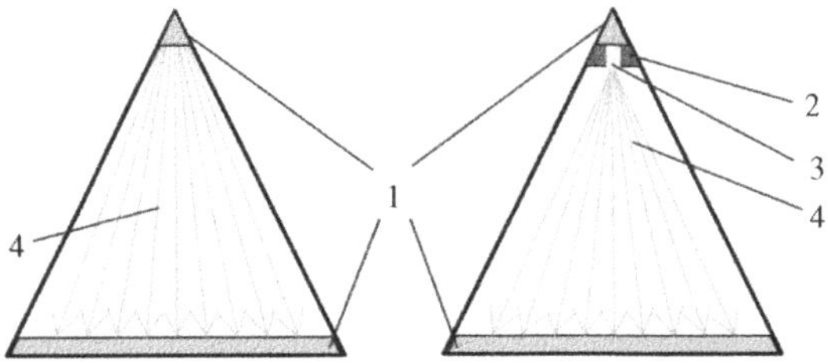

Figure 12.11 Flow path of surface leakage current on the triangular glass plate: (a) without dry band; (b) with dry band and partial electric arc. 1: Electrode; 2: dry band; 3: partial electric arc; 4: current path. *Source*: Cao et al. (2018)/with permission of IEEE.

According to Eq. (12.1), such a change results in a reduction of the equivalent diameter $D(x)$, which in turn increases the effective form factor of the current path. Consequently, if the original form factor—defined under ideal, arc-free conditions—is used in the calculation, the surface conductivity deduced from the leakage current will appear lower than the actual value. This underestimation arises because the modified current path, shaped by arc formation, does not conform to the assumptions of uniform geometry and uninterrupted surface conduction.

For the triangular glass plate, when the dry band formed on the surface of the glass plate, along with a partial electric arc across the dry band, the path of leakage current did not change much, as is shown in Figure 12.11. It could be found that there was little difference between the surface conductivity calculated using the maximum value of the leakage current during the wetting process and the theoretical value.

12.5 Influence of Partial Electric Arc on the Contamination Monitoring in View of Leakage Current

The leakage current and voltage distribution along the surface of the insulator can be effectively analyzed using numerical simulation tools. In this study, the commercial finite element analysis software COMSOL Multiphysics 5.0 was employed to model and solve the surface conduction problem using the Finite Element Method (FEM).

A detailed model of the XWP-70 porcelain insulator was constructed to simulate the surface current distribution under conditions where dry bands form and are partially bridged by electric arcs. The simulation utilized COMSOL's Electric

Currents physics interface in conjunction with the Shell feature, allowing accurate representation of current flow on the curved insulator surface. Relevant theoretical and technical documentation supporting this method can be found in the official COMSOL help resources.

To simplify the arc modeling in the simulation, the partial electric arc was represented using a copper conductor, which provided a low-resistance path across the dry band. The surface conductivity of the contamination layer under saturated moisture conditions was set to 100 μS, while the conductivity of the dry band region was reduced to 0.001 μS, reflecting the near-insulating behavior observed in experiments.

Based on experimental observations presented earlier in this chapter, the arc cross-connection length—representing the effective leakage path—ranged between 2 and 6 cm. The complete simulation setup, including geometry and boundary conditions, is illustrated in Figure 12.12.

The effective radius of the partial arc root, which strongly influences current distribution, is known to depend on the leakage current magnitude. This relationship was modeled according to Wilkins' empirical result, which is expressed in Eq. (12.4):

$$R_0 = \sqrt{\frac{I}{1.45\pi}} \tag{12.4}$$

Because the test result showed that the leakage was in a range of 153–1130 mA, the effective radius of the arc root was set as 0.37 cm correspondingly in this chapter.

9 kV was applied at both the steel cap and pin of the insulator and the potential voltage distribution on the surface with/without arc is shown in Figure 12.13, where the length of partial arc was 4 cm. Then the surface current on the insulator in the two cases was calculated as follows:

$$I_1 = 0.815 \text{ A}$$
$$I_2 = 0.611 \text{ A}$$

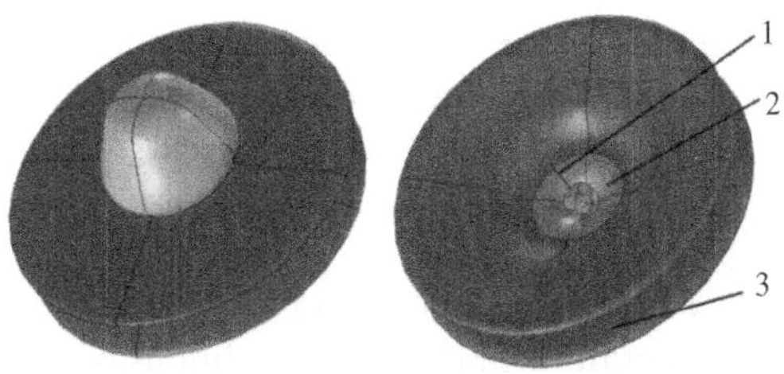

Figure 12.12 Calculation models of XWP-70. 1: partial arc; 2: dry band; 3: contamination layer. *Source*: Cao et al. (2018)/with permission of IEEE.

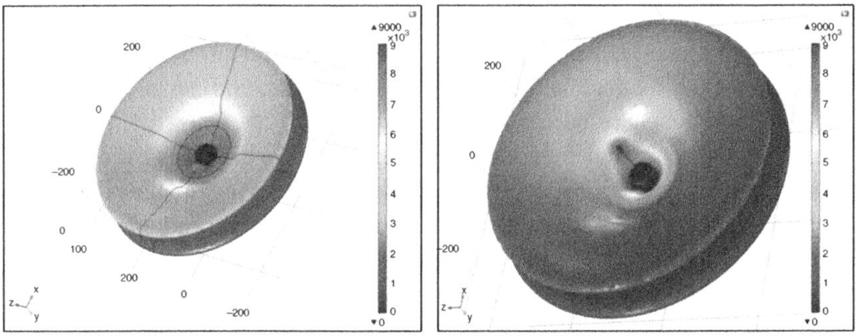

Figure 12.13 Potential electric distribution of the insulator: (a) without dry band; (b) with dry band and partial electric arc. *Source*: Cao et al. (2018)/with permission of IEEE.

where I_1 is the leakage current without the dry band and I_2 is that with the dry band as well as the electric arc path. The surface conductivity of pollution layers can be obtained using Eq. (12.2).

$$K_1 = 100.02\ \mu S$$

$$K_2 = 74.95\ \mu S$$

It can be found that, for XWP-70 insulators, when the dry band and partial electric arc formed, the leakage current reflected the total resistance of the residual contamination layers and the partial electric arc. Compared with the former, the electric arc resistance was very almost negligible. However, due to the small radius of partial electric arc root, the contact area between it and the contamination layer was very small, resulting in the uneven current distribution on the insulator surface, which was the exact reason why Eq. (12.2) was not applicable for the calculation of surface conductivity using the leakage current.

It was found that, when both the SDD of the insulator and the effective radius of the electric arc root remained constant, variations in the width of the dry bands on the insulator surface—or in the leakage distance of the partial electric arc crossing the insulator—had negligible influence on the calculated surface conductivity derived from the leakage current, within a certain range. This result suggests that the effect of these geometrical variations on leakage current is limited under fixed arc root and contamination conditions. The relationship is illustrated in Figure 12.14.

Based on the preceding analysis, it can be concluded that the theoretical surface conductivity of an insulator under saturated moisture conditions is determined solely by the amount of dissolved soluble salt in the contamination

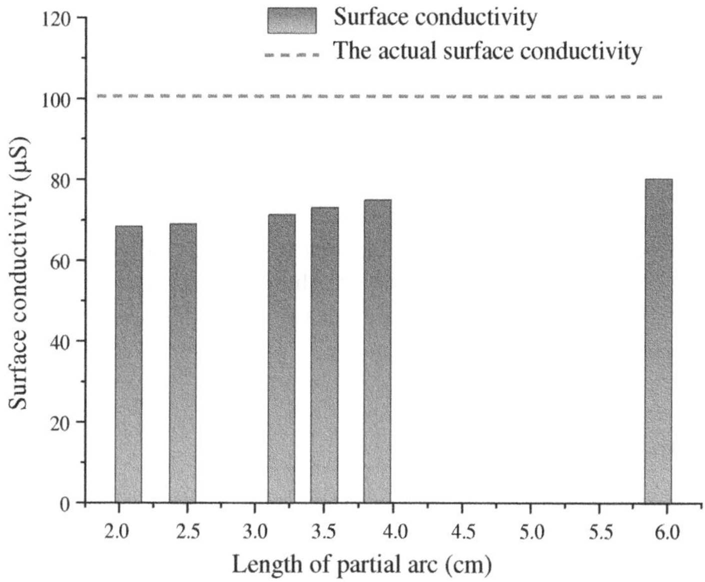

Figure 12.14 Electric field distribution of the insulator.

layer. For artificially applied pollution, where NaCl is used as the soluble salt, a clear linear relationship exists between surface conductivity and the SDD. In such cases, the theoretical ratio of surface conductivity to SDD is approximately 1,000.

However, under real-world operating conditions, the surface conductivity values calculated using Eq. (12.2)—based on maximum leakage current measurements—deviate from this theoretical expectation. For example, in the case of the XWP-70 insulator, the measured ratio falls within the range of 750–850, while for the XP-70 insulator, it is significantly lower, ranging from 350–550.

This discrepancy arises because, during the wetting process under energized conditions, dry bands and partial electric arcs tend to form around the steel pin region of the insulator. These phenomena alter the leakage current path, causing it to reflect not only the conductivity of the residual wet contamination but also the combined resistance of both the partial arc path and the dry contamination layer. While the resistance of the electric arc is relatively low—and often considered negligible—it can still influence the effective current path geometry, thereby affecting the calculation of surface conductivity based on leakage current.

Notably, the presence of the partial arc effectively alters the form factor of the insulator. As a result, the derived surface conductivity no longer strictly corresponds to the actual conductivity of the contamination layer. However, for a given insulator type, if the arc length remains within a predictable range, the calculated surface conductivity remains proportional to the true surface conductivity of the pollution layer.

It is important to emphasize that the extent of this effect varies between insulator designs, depending on differences in geometry, electric field distribution, and the location and frequency of arc formation. These factors must be carefully considered when interpreting leakage current measurements for contamination assessment in real-world monitoring applications.

12.6 Summary

The theoretical surface conductivity of the insulator under the saturated moisture condition is independent of measurement methods and it is only related to the amount of soluble salts dissolved in the contamination. However, it is found that the surface conductivity deduced from the leakage current is different from the theoretical surface conductivity.

The main cause is that, during the wetting process of the insulator in actual operation, a dry band and partial electric arc will be generated around the steel pin. Although the electric arc resistance is very small and even negligible, the path of the current on the residual contamination layer is changed, resulting in the uneven distribution of the leakage current on the insulator surface.

The investigation results have shown that, the surface conductivity converted from the leakage current is in proportion to the actual surface conductivity of the contaminated layer. However, for insulators with different profiles, the influence of partial electric arc can be different. The ratio of surface conductivity of the XWP-70 insulator under the saturated moisture condition to SDD is in a range of 750–850, while that of XP-70 insulator is roughly between 350 and 550.

Bibliography

Amarh, F., Karady, G.G., and Sundararajan, R. Linear stochastic analysis of polluted insulator leakage current, IEEE Trans. Power Del., 2002, 17 (4), 1063–1069.

Cao, B., Wang, L., Li, X. et al. Guan, Influence of partial electric arc on the contamination degree monitoring system in view of leakage current, IEEE Trans. Dielectr. Electr. Insul., 2018, 25 (4), 1545–1552. https://doi.org/10.1109/TDEI.2018.006944.

Chrzan, K. Conductivity of aqueous dust solutions, IEEE Trans. Electr. Insul., 1987, 22, 241–244.

Dong, H., Xu, W., Cao, B. et al. Effect of soluble salt on conductivity of partial surface, Annual Report IEEE Conference on Electrical Insulation and Dielectric Phenomena (CEIDP), 2015, pp. 185–188.

Farzaneh, M. and Chisholm, W. Insulators for Icing and Polluted Environments, Wiley, 2009.

Fierro-Chavez, J.L., Ramirez-Vazquez, Z., and Montoya-Tena, G. On-line leakage current monitoring of 400 kV insulator strings in polluted areas, IEE Proc. Gener. Transm. Distrib., 1996, 143 (6), 560–564.

Holtzhausen, J.P. AC pollution flashover models: accuracy in predicting insulator flashover. Proceedings of the 10th International Symposium on High Voltage Engineering, 1997.

Holtzhausen, J.P. and Vosloo, W.L. The pollution flashover of AC energized post type insulators: the role of shortening of the arc, IEEE Trans. Dielectr. Electr. Insul., 2001, 8 (2), 191–194.

IEC 60507-2013, Artificial Pollution Tests on High Voltage Insulators to be Used on AC systems, 2013.

IEC/TS 60815-1, Selection and Dimensioning of High-voltage Insulators Intended for Use in Polluted Conditions-Part 1: Definitions, Information and General Principles, 2008.

IEEE Std. 4-2013, IEEE Standard Techniques for High-Voltage Testing, IEEE Publication, 2013.

Jiang, X., Yan, S., Sun, C. et al. Evaluating the safety condition of porcelain insulators by the time and frequency characteristics of LC based on artificial pollution tests, IEEE Trans. Dielectr. Electr. Insul., 2010, 17 (2), 481–489.

Li, J., Guan, Z., Wang, L. et al. An experimental study of AC arc propagation over a contaminated surface, IEEE Trans. Dielectr. Electr. Insul., 2012, 19 (4), 1360–1368.

Li, J., Sun, C., Sima, W. et al. Contamination level prediction of insulators based on the characteristics of leakage current, IEEE Trans. Power Del., 2010, 25 (1), 417–424.

Rizk, F.A.M. A criterion for AC Flashover of Polluted Insulators, IEEE Power Engineering Society Winter Power Meeting, New York, NY, USA, 1971.

Takasu, K., Shindo, T., and Arai, N. Natural contamination test of insulators with DC voltage energization at inland areas, IEEE Trans. Power Del., 1988, 3 (4), 1847–1853.

Wang, L., Cao, B., Mei, H. et al. Effects of natural contamination components on the surface conductivity under saturated moisture, IEEE Trans. Dielectr. Electr. Insul., 2017, 24 (5), 2945–2951.

Wilkins, R. Flashover voltage of high-voltage insulators with uniform surface-pollution films, Proc. IEE, 1969, 116, 457–465.

Zhang, Z., Zhang, D., Jiang, X. et al. Study on natural contamination performance of typical types of insulators, IEEE Trans. Dielectr. Electr. Insul., 2014, 21 (4), 1901–1909.

Zhao, C.L., Mei, H.W., Zhu, M.X. et al. Development of contamination flashover pre-warning system and analysis of operating experience, IEEE Trans. Electr. Insul., 2015, 22 (4), 2234–2241.

13

Spatial and Temporal Analysis of Contamination on Insulator Surfaces

13.1 Introduction

The final technical section (Chapters 13–14) presents advanced analysis and correction strategies for soluble salt contamination on insulation surfaces. It includes spatial and temporal analysis methods and solutions to improve contamination assessments under changing environmental condition.

First, to evaluate the degree of pollution on insulator surfaces and support the external insulation design of power transmission and transformation equipment, the International Electrotechnical Commission (IEC) recommends several indices—such as Equivalent Salt Deposit Density (ESDD), Site Equivalent Salinity, and Dust Deposit Gauge Index (DDGI)—to characterize pollution levels on insulators.

However, ESDD has limitations. For instance, insulators with the same ESDD can exhibit widely scattered pollution flashover voltages. Moreover, the flashover voltage of naturally contaminated insulators sometimes differs from that of artificially contaminated insulators, even when both have the same Salt Deposit Density (SDD). In artificial contamination tests, sodium chloride (NaCl) is commonly used to simulate the soluble salt components found in natural pollution. Due to its high solubility and fast dissolution rate, NaCl fully dissolves when the insulator surface reaches moisture saturation. As a result, in artificial tests, the flashover voltage of polluted insulators tends to show a negative exponential relationship with SDD and exhibits relatively low result dispersion.

In real-world conditions, pollution sources along transmission lines are typically diverse. According to CIGRE Task Force 33.04.01, typical pollution environments include marine, industrial, agricultural, and desert areas. These varied sources—such as seawater spray, road de-icing salts, or cooling tower emissions—can deposit different types of salts on the insulator surface, including highly soluble salts like NaCl, KCl, and Na_2SO_4, as well as slightly soluble ones such as $CaSO_4$.

The measurement of ESDD involves significant water consumption, up to 200 mg/cm^2, as stipulated by standards like IEC 60815-1. In contrast, the actual water present on outdoor insulators under saturated moisture conditions is typically less than 5 mg/cm^2. This discrepancy means that slightly soluble salts, which are fully dissolved during ESDD testing, may only partially dissolve in real conditions. Therefore, even with the same ESDD, the flashover voltage of naturally contaminated insulators can vary significantly depending on the region.

Accurately measuring the natural contamination composition on insulator surfaces is crucial for creating reliable pollution maps and guiding external insulation design. In this study, contamination deposited on specimens from various regions and under different installation conditions was analyzed. The effects of inclination angle and deposition duration were examined. Additionally, to investigate the influence of insulator material, shed profile, and voltage polarity, long-term operational contamination on post insulators was assessed. Since pollution sources play a critical role in contamination deposition, insulators located near different types of sources were also studied. The findings aim to provide valuable references for the design and selection of external insulation in power systems.

13.2 Artificial Contamination Test

13.2.1 Test Specimens, Facilities and Procedure

Ideally, the most accurate way to study natural contamination on insulator surfaces is by periodically removing industrial insulators from operating transmission lines and analyzing their contamination components. However, this approach is impractical due to the high cost and disruption to power system operation. As a viable alternative, contamination can be studied using test specimens exposed to the atmosphere under non-energized conditions.

To investigate the variation of pollution components over time, a simple natural contamination deposition platform was constructed, as illustrated in Figure 13.1. A total of 90 glass specimens, each measuring 3 cm × 5 cm × 0.5 cm (length × width × height), were prepared for the experiment. Before exposure, all specimens were thoroughly cleaned to ensure they were free from any pre-existing contaminants. Following standard procedures, each specimen was washed with a slurry of distilled water and kaolin, then rinsed with distilled water to remove any remaining dirt or grease. A specimen was considered sufficiently clean if large, continuous wet areas appeared during rinsing, indicating the absence of grease.

(a) (b)

Figure 13.1 Test platform: (a) pollution deposition test platform; (b) test specimens. *Source*: Cao et al. (2020)/with permission of John Wiley & Sons.

(a) (b)

Figure 13.2 Analysis equipment: (a) ion chromatography instrumentation (ICS 1500); (b) inductively coupled plasma optical emission spectrometer (Perkin Elmer Optima 7300 DV). *Source*: Cao et al. (2020)/with permission of John Wiley & Sons.

To prevent rainfall from influencing the deposition process, the cleaned specimens were carefully handled and placed into a specially designed louvre box positioned three meters above the ground. This box allowed exposure to atmospheric contaminants while shielding the specimens from direct precipitation.

The test lasted 21 days. Every week, five specimens from each inclination angle—horizontal (0°), 45°, and vertical (90°)—were retrieved from the louvre box for contamination analysis. To evaluate the composition of soluble salts, the naturally deposited contamination on the specimen surfaces was analyzed using Inductively Coupled Plasma Optical Emission Spectroscopy (ICP-OES) and ion chromatography, as depicted in Figure 13.2. Each sample was analyzed once using ICP-OES. For every orientation group, the results from five specimens were averaged to ensure statistical reliability.

Inductively Coupled Plasma Optical Emission Spectroscopy (ICP-OES) is a rapid, multi-element analytical technique that offers a wide dynamic linear range and moderate detection limits (typically 0.2–100 ppb). The technique utilizes an inductively coupled plasma source generated by ionizing argon gas, which dissociates the sample into its constituent atoms or ions and excites them to emit light at characteristic wavelengths. Up to 60 elements can be detected in a single sample run, typically completed in under one minute.

IC, commonly employed in water quality analysis, enables detection of major ion species in the parts-per-billion (ppb) range. In this method, sample solutions are pumped through an ion exchange column, which exhibits varying affinities for different ions. This differential interaction causes the ions to elute at different rates. As the ions elute, they are detected through changes in the conductivity of the mobile phase. IC can measure major anions such as fluoride, chloride, nitrate, nitrite, and sulfate, as well as major cations including lithium, sodium, ammonium, potassium, calcium, and magnesium. It is also capable of detecting organic acids.

To analyze the anionic and cationic composition of the contamination in this study, the following procedure was used:

1. **Collection**: Contaminants deposited on the surface of each specimen were gently scrubbed using a clean cloth specifically designated for contamination collection.
2. **Dissolution and Filtration**: The collected material was dissolved in a known volume of deionized water and then filtered multiple times through 0.45 μm (or finer) filters to produce a clear contamination solution.
3. **Ion Chromatography Analysis**: The resulting solution was diluted 10-fold and analyzed using a Dionex ICS-1500 ion chromatography system (USA). Ion concentrations were determined by calculating the area under each chromatographic peak, where a larger peak area corresponds to a higher concentration of a specific ion species.
4. **ICP-OES Analysis**: Simultaneously, an aliquot of the same contamination solution was analyzed using ICP-OES to identify and quantify the cationic components present.

This dual-method approach ensures comprehensive and accurate profiling of the ionic contaminants naturally deposited on insulator surfaces.

13.2.2 Tests at Different Contamination Durations in the Same Area

To study the temporal variation of contamination, all specimens were exposed to natural atmospheric conditions for a total duration of three weeks. Every seven

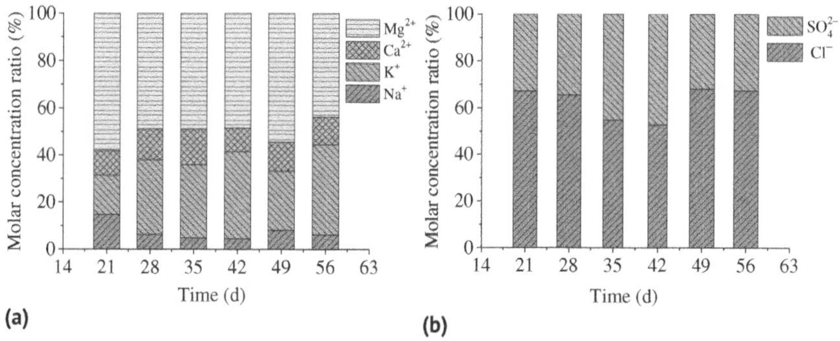

Figure 13.3 Changes of soluble salt composition of contamination with time (horizontal arrangement): (a) proportion of molar concentration of different cations; (b) proportion of molar concentration of different anions. *Source*: Adapted from Cao et al. (2020).

days, five horizontally arranged specimens were retrieved and sent to the laboratory for testing. The concentrations and proportions of soluble salts deposited on the glass surfaces over time were analyzed, and the results are presented in Figure 13.3.

By comparing the anionic and cationic compositions of the contaminants at each sampling interval, it was observed that the relative proportions of the major ions remained relatively stable throughout the three-week exposure period. This indicates that while the total quantity of deposition may vary with time, the composition of the soluble salts accreted on the specimen surface exhibits minimal fluctuation under the same environmental conditions.

13.2.3 Tests at Different Areas

To investigate the influence of geographical location on contamination composition, natural contamination tests were conducted in two distinct areas: Shenzhen, Guangdong Province, and Nanjing, Jiangsu Province, using the same test platform described in Section 13.2.1. To further analyze the source of soluble salt pollutants, a soluble salt collection device was designed and deployed near each test site, as illustrated in Figure 13.4.

This device was installed adjacent to the contamination test platform to capture airborne soluble pollutants from the ambient atmosphere. During operation, the device utilized an air pump calibrated to an inlet flow rate of 2.7 L/min. The drawn air was passed into a sealed measuring cylinder

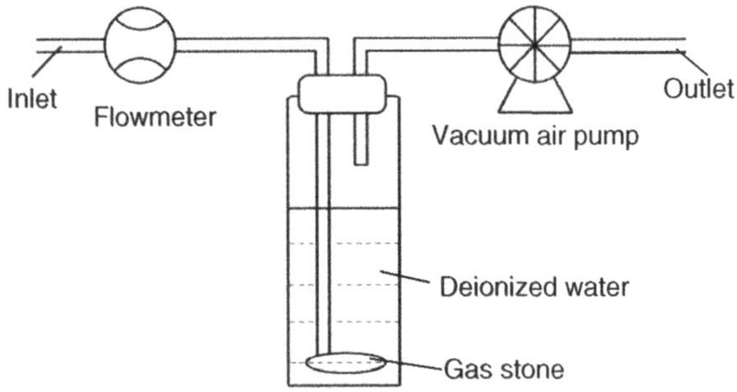

Figure 13.4 Soluble salt collection device for atmospheric sampling.
Source: Cao et al. (2020)/with permission of John Wiley & Sons.

containing 50 mL of deionized water. To maximize the contact between air and water, an air stone (commonly known as an aquarium bubbler) was placed at the bottom of the cylinder. This allowed fine air bubbles to disperse evenly, promoting efficient dissolution of airborne soluble salts into the water.

The pump operated in cycles—running for 15 minutes followed by a 45-minute pause. After the test period, water samples were collected from the device and analyzed to determine the molar proportions of soluble anions and cations. These tests were conducted concurrently in Shenzhen and Nanjing.

After one month of natural contamination exposure, the molar concentration proportions of soluble ions in both the deposited contamination (on test specimens arranged horizontally, at 45°, and vertically) and in the ambient atmosphere were measured. The results are presented in Figures 13.5 and 13.6.

As shown in Figures 13.5 and 13.6, the arrangement of specimens—horizontal, inclined at 45°, or vertical—had minimal effect on the composition of soluble salts in the contamination. Additionally, the composition of soluble ions found on the specimen surfaces was found to closely resemble the ion proportions measured in the surrounding atmosphere, indicating that the deposited contamination directly reflects local airborne pollutants.

However, a clear difference was observed between the two locations. The relative proportions of anions and cations in the contamination varied significantly between Shenzhen and Nanjing, reflecting differences in regional atmospheric pollution sources. These findings reinforce the importance of local environmental conditions in determining the chemical characteristics of insulator surface contamination.

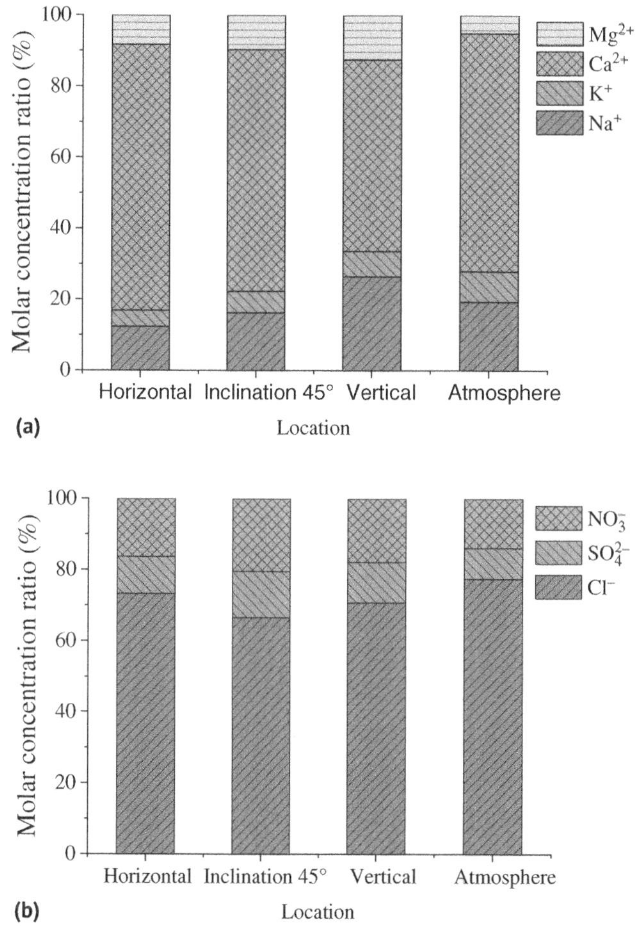

Figure 13.5 Contamination of naturally contaminated test samples (Shenzhen, Guangdong Province): (a) proportion of molar concentration of cations; (b) proportion of molar concentration of anions. *Source*: Adapted from Cao et al. (2020).

13.3 Contamination Components of Actual Operating Insulators

In addition to laboratory-exposed specimens, studying contamination on insulators that have been in long-term operation provides more realistic insights into pollution accumulation under actual service conditions. These data are essential for validating lab-based simulations and improving external insulation design.

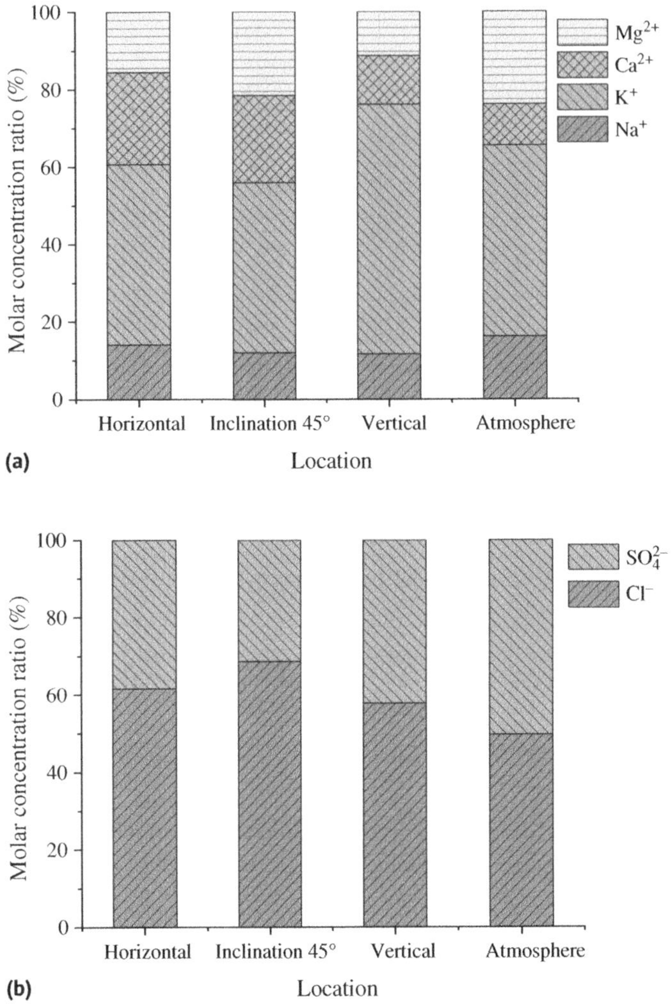

Figure 13.6 Contamination of naturally contaminated test samples (Nanjing, Jiangsu Province): (a) proportion of molar concentration of cations; (b) proportion of molar concentration of anions. *Source:* Adapted from Cao et al. (2020).

13.3.1 Test Specimens, Facilities, and Procedure

To assess the contamination components on in-service insulators, post insulators that had been in continuous operation were retrieved from substations for analysis. The selected insulators represented typical installations and had been exposed to natural environmental conditions without cleaning or maintenance for extended periods.

Upon removal, each insulator was labeled, photographed, and stored under controlled conditions to avoid post-removal contamination or alteration. Specific attention was given to:

1. Material type (glass, porcelain, or composite),
2. Shed profile (standard, aerodynamic, or anti-pollution types),
3. Mounting orientation (vertical or inclined),
4. Voltage polarity during service (AC or DC systems).

To collect contamination samples, each insulator was carefully cleaned section by section using deionized water and soft brushes. The wash water was filtered and processed similarly to the lab specimen protocols and analyzed via IC and ICP-OES to determine soluble anion and cation concentrations. This ensured consistency and comparability with the earlier laboratory studies.

13.3.2 Tests at the Same Substation (Same Area)

During a scheduled power-off overhaul of the Tianshengqiao–Guangzhou ±500 kV DC transmission project, an on-site contamination analysis was conducted at the Guangzhou converter station. The study focused on six post insulators in service, as shown in Figure 13.7. These insulators were selected to represent different installation conditions and material configurations within the

Figure 13.7 Photograph of on-site equipment at Guangzhou station. *Source*: Cao et al. (2020)/with permission of John Wiley & Sons.

Table 13.1 Post insulators at Guangzhou converter station.

No.	Post insulator	Sheds	Material
#1	+500 kV DC	Uniform	RTV coated
#2	+500 kV DC	Alternating	HTV coated
#3	+500 kV DC	Alternating	RTV coated
#4	−500 kV DC	Alternating	RTV coated
#5	220 kV AC	Alternating	Porcelain
#6	neutral line	Uniform	Porcelain

same substation. The shed types, surface materials, and operating voltage types (AC, DC, or neutral) for each of the six insulators are detailed in Table 13.1.

For contamination sampling, each shed unit was treated as a single entity—no distinction was made between the upper and lower surfaces. This approach aimed to simplify the sampling process while capturing the cumulative contamination present on the insulator surfaces.

The collected contamination was dissolved, filtered, and analyzed using Inductively Coupled Plasma Optical Emission Spectroscopy (ICP-OES) to identify and quantify the soluble salt components. Each insulator was measured once. The results provide insight into how surface material, shed geometry, and voltage polarity influence the accumulation of ionic contaminants under real operating conditions.

The measured proportions of soluble anions and cations in the surface contamination of the six post insulators are presented in Figure 13.8. Among them, Post Insulator #1 features uniform sheds, while Post Insulator #2 has alternating shed geometry. Despite the structural difference, both insulators exhibit nearly identical soluble ion compositions in their surface contamination.

As summarized in Table 13.1, Post Insulator #3, #4, and #5 were operated under +500 kV DC, −500 kV DC, and AC 220 kV, respectively. Post Insulator #6 was in a de-energized (neutral) state. Nevertheless, the proportions of soluble salts found on the surfaces of all six insulators remained remarkably consistent, regardless of their voltage polarity or energization status.

In terms of surface material, Post Insulators #1, #3, and #4 were coated with RTV (Room Temperature Vulcanized silicone rubber). Post Insulator #2 was made of HTV (High Temperature Vulcanized silicone rubber), and Post Insulator #5 was composed of porcelain. Despite the variation in surface materials, the composition of soluble salt ions in their respective contaminations showed minimal deviation.

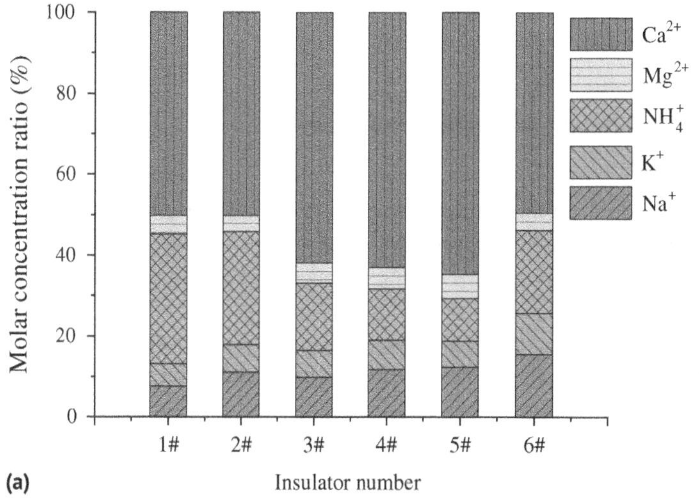

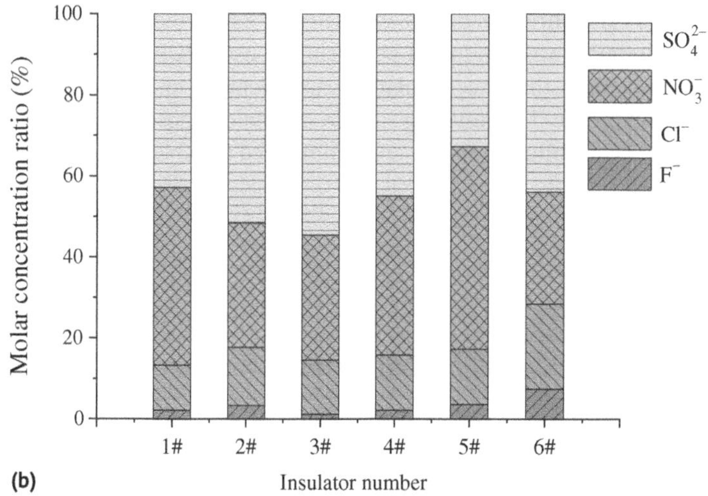

Figure 13.8 Contamination of naturally contaminated post insulators at Guangzhou station: (a) proportion of molar concentration of cations; (b) proportion of molar concentration of anions. *Source*: Adapted from Cao et al. (2020).

These findings indicate that for post insulators located within the same substation and subjected to the same atmospheric environment, the proportions of soluble anions and cations in the contamination are nearly identical. Furthermore, the data suggest that the shed geometry, voltage polarity, and surface material of the insulators do not significantly affect the relative composition of soluble salts deposited on their surfaces.

13.3.3 Tests at Different Areas (Near the Typical Pollution Sources)

Given that contamination on insulator surfaces primarily originates from airborne pollutants in the surrounding atmosphere, this section examines the influence of local pollution sources on the composition of soluble salts deposited on insulators.

To conduct this analysis, insulators were collected from transmission lines or substations located within a 3 km radius of typical pollution sources. These sources were classified into six representative categories based on their emission characteristics: power plants, cement plants, chemical plants, agriculture, coastal areas, and salt lakes, as shown in Figures 13.9 and 13.10. Each pollution category is generally associated with characteristic types of particulate and gaseous emissions. These are summarized in Table 13.2.

For each pollution site, three insulators were selected for testing. A single contamination measurement was conducted per insulator using ICP-OES. The ion concentration results from the three insulators were averaged to represent each site's contamination profile. The resulting soluble salt ion compositions for insulators exposed near each pollution source are summarized in Table 13.3.

Figure 13.9 Typical sources of pollution environment caused by human activity: (a) power plants; (b) cement plants; (c) chemical plants; (d) agriculture. *Source*: Cao et al. (2020)/with permission of John Wiley & Sons.

(a)

(b)

Figure 13.10 Typical sources of natural pollution environment: (a) coastal areas; (b) salt lake area. *Source*: Cao et al. (2020)/with permission of John Wiley & Sons.

Table 13.2 Typical pollution sources.

No.	Typical pollution source	Specific characteristics
(a)	Power plants	Main characteristic pollutants are gaseous pollutants such as sulphur dioxide and nitrogen oxides, as well as a certain amount of solid pollutants such as coal powder and coal ash.
(b)	Cement plants	Pollution of insulators contains a large amount of cement dust.
(c)	Chemical plants	Pollution of the insulators are subject to raw materials or products from chemical plants, and the characteristic effluent of different chemical plants are different.
(d)	Agricultural pollution	Pollution of the insulators contain a certain amount of pesticide fertilizers or crops and their burning residue.
(e)	Coastal areas	Coastal areas are generally relatively open. In addition to the influence of typical artificial pollution sources (factories) in the surrounding area, coastal transmission lines are also affected by the salt spray.
(f)	Salt lakes	The water vapor surrounding this pollution source has a high salt content.

By comparing the soluble salt ion compositions in the surface contamination of insulators located near the six typical pollution sources—summarized in Table 13.3—several key observations can be made:

1. Significant variation exists among different pollution categories: Each pollution source type exhibits a distinct profile of soluble salt contamination. For example, calcium sulfate ($CaSO_4$) content is notably low in coastal environments, where sodium chloride (NaCl) dominates due to marine aerosol

Table 13.3 Soluble salt ion composition in surface contamination of insulators.

	Molar concentration ratio of anions (%)			Molar concentration ratio of cations (%)				
Pollution sites	Cl^-	NO_3^-	SO_4^{2-}	Na^+	K^+	Mg^{2+}	Ca^{2+}	NH_4^+
Power station #1	9.2	12.0	78.8	5.6	3.7	4.4	79.3	7.0
Power station #2	37.2	15.0	47.9	19.5	5.2	1.9	73.4	0.0
Cement plant #3	11.2	15.0	73.7	6.1	2.5	3.0	76.1	12.3
Cement plant #4	8.4	13.8	77.8	4.8	3.2	3.6	78.4	10.1
Cement plant #5	1.7	1.7	96.6	1.7	1.7	1.2	92.1	3.2
Chemical plant #6	36.6	9.7	53.7	59.7	14.5	2.8	23.0	0.0
Chemical plant #7	50.4	2.4	47.2	49.1	8.9	4.5	37.6	0.0
Chemical plant #8	9.0	28.2	62.9	10.3	4.0	7.2	76.2	2.2
Farmland #9	8.5	12.3	79.1	7.3	3.5	2.9	74.2	12.1
Farmland #10	11.1	17.6	71.2	8.9	3.8	3.8	77.7	5.8
Farmland #11	12.3	19.5	68.3	8.0	2.8	3.4	54.7	31.1
Coastal areas #12	2.3	82.4	15.3	13.2	5.1	79.5	2.2	0.0
Coastal areas #13	61.5	10.9	27.5	65.9	4.8	9.1	20.2	0.0
Salt Lake #14	49.8	9.0	41.2	17.2	1.5	74.9	6.4	0.0
Salt Lake #15	70.8	4.0	25.2	33.8	1.9	16.6	47.8	0.0
Salt Lake #16	68.0	4.1	27.9	23.3	5.6	26.4	44.8	0.0

influence. In contrast, inland industrial areas, such as those near cement and power plants, show much higher $CaSO_4$ concentrations. These findings are consistent with previous studies and reflect the nature of local particulate and gaseous emissions.

2. Intra-category variability across different sites can be substantial: While similar pollution sources can produce comparable contamination in some cases, the composition of soluble salts on insulator surfaces can also vary significantly within the same category, depending on local environmental factors. For example, the proportion of $CaSO_4$ in the contamination of insulators near cement plant #5 is more than 92%, while the proportion of $CaSO_4$ in the surface contamination of naturally contaminated insulators around cement plant #5 is less than 74%. Moreover, insulators around farmland #9, #10 and #11 have highly different soluble salt composition in the contamination. The Ca^{2+} accounts for 74.2% of the cations in the contamination solution of insulator

around farmland #9, while the Ca^{2+} in the contamination solution of surface contamination on the insulator around farmland #11 accounts for only 54.7% of the total cations, and NH_4^+ accounts for 31.1% of the total cations.

These findings highlight the complex interplay between local pollutant sources and atmospheric transport conditions. They also emphasize the necessity of site-specific contamination assessments for effective insulation design and pollution monitoring in power systems.

13.4 Summary

The spatial and temporal distribution characteristics of soluble salt compositions on naturally polluted insulators were systematically investigated in this study. Based on the experimental results and analysis, the following conclusions can be drawn:

1. The composition of soluble salts on the surface of naturally contaminated specimens remained stable over time within the same geographic location. No significant variation in the proportions of anions and cations was observed with increased exposure duration.
2. The orientation of the specimens (horizontal, inclined, or vertical) had no noticeable impact on the composition of the deposited soluble salts. Furthermore, the ion composition of surface contamination was consistent with that of the surrounding atmospheric environment, indicating direct deposition from airborne pollutants.
3. For in-service insulators operating in the same substation or geographic area, the soluble salt composition was similar across units, regardless of differences in shed geometry, surface material (e.g., RTV, HTV, porcelain), voltage polarity (AC/DC/neutral), ground height, or pollution severity.
4. In contrast, substantial differences were found in the composition of soluble salts on insulators located near different types of pollution sources (e.g., coastal, agricultural, industrial). Even within the same category of pollution source, contamination composition may vary considerably from one site to another, influenced by localized environmental and operational factors.

Bibliography

Ahmadi-Joneidi, I., Majzoobi, A., and Shayegani-Akmal, A.A. Aging evaluation of silicone rubber insulators using leakage current and flashover voltage analysis, IEEE Trans. Dielectr. Electr. Insul., 2013, 20 (1), 212–220.

Cao, B. and Wang, L. Wetting characteristics of artificial contamination on the hydrophobic surface. Eng. Fail. Anal., 2019, 100, 428–438.

Cao, B., Zhao, C., Yin, F. et al. Temporal and spatial characteristics of soluble salt components accreted on insulator surface, IET Sci. Meas. Technol., 2020, 14 (8), 891–896.

CIGRE Task Force 33.04.01. Polluted Insulators: A Review of Current Knowledge, CIGRE Brocure, No. 158, June, 2000.

Deb, S., Choudhury, N.R., Ghosh, R. et al. Short time modified Hilbert transform-aided sparse representation for sensing of overhead line insulator contamination, IEEE Sens. J., 2018, 18 (19), 8125–8132.

Farzaneh, M. and Chisholm, A.W. Insulators for Icing and Polluted Environments, Wiley Press, 2009.

Gençoğlu, M.T. and Cebeci, M. The pollution flashover on high voltage insulators, Electr. Power Syst. Res., 2008, 78 (11), 1914–1921.

IEC 60815-3-2008, Selection and Dimensioning of High-voltage Insulators Intended for Use in Polluted Conditions - Part 3: Polymer Insulators for A.C. Systems, 2008.

IEC/TS 60815-1-2008, Selection and Dimensioning of High-voltage Insulators Intended for Use in Polluted Conditions - Part 1: Definitions, Information and General Principles, 2008.

Jin, L., Tian, Z., Ai, J. et al. Condition evaluation of the contaminated insulators by visible light images assisted with infrared information, IEEE Trans. Instrum. Meas., 2018, 67 (6), 1349–1358.

Li, L., Jiang, Y., Lu, M. et al. Study on the difference of chemical composition of insulator contamination on UHV-AC and -DC transmission lines, IET Sci. Meas. Technol., 2018, 12 (1), 17–24.

Liu, Y., Du, B., and Farzaneh, M. Self-normalizing multivariate analysis of polymer insulator leakage current under severe fog conditions, IEEE Trans. Power Del., 2017, 32 (3), 1279–1286.

Punse, G.S. Analysis and design of transmission tower, Int. J. Mod. Eng. Res. (IJMER), 2014, 4 (1), 116–138.

Pylarinos, D., Siderakis, K., Mavrikakis, N. et al. Mapping HV insulators pollution in the Mediterranean island of Crete, International Conference on Deregulated Electricity Market Issues in South Eastern Europe, Cyprus, September 2014, pp. 1–5.

Takasu, K., Shindo, T., and Arai, N. Natural contamination test of insulators with DC voltage energization at inland areas, IEEE Trans. Power Del., 1988, 3 (4), 1847–1853.

Velásquez, R.M., Lara, J.V.M., The need of creating a new nominal creepage distance in accordance with heaviest pollution 500kV overhead line insulators, Eng. Fail. Anal., 2018, 86, 21–32.

Verma, A.R. and Subba, R.B. Interpretation of surface degradation on polymeric insulators, Eng. Fail. Anal., 2019, 95, 214–225.

Wang, L., Cao, B., Mei, H. et al. Effects of natural contamination components on the surface conductivity under saturated moisture, IEEE Trans. Dielectr. Electr. Insul., 2017, 24 (5), 2945–2951.

Wu, W.N., Wu, G.Y., and Zhang, Q. Application and operation characteristics of 1000kV UHV AC insulator, High volt. Technol., 2011, 37, 15–20.

Yin, F., Jiang, X., and Farzaneh, M. Electrical performance of 330-kV composite insulators with different shed configurations under icing conditions, IEEE Trans. Dielectr. Electr. Insul., 2015, 22 (6), 3395–3404.

14

Correction Methods for Soluble Salt Components in Contamination Assessment

14.1 Introduction

As discussed in the previous chapter, the soluble salt content of insulator surface contamination is influenced by local environmental pollution. Naturally deposited contaminants vary significantly with local sources, containing a mixture of highly soluble salts (e.g., NaCl, KCl, and Na_2SO_4) and slightly soluble salts (e.g., $CaSO_4$), the latter of which dissolve only partially under limited moisture conditions. In contrast, the standard Equivalent Salt Deposit Density (ESDD) measurement procedure employs excess water, resulting in complete dissolution of even slightly soluble salts. Consequently, the measured ESDD may not accurately represent the actual electrical behavior of natural contamination under typical environmental conditions. If this uncorrected ESDD is used as the basis for artificial pollution tests, the resulting flashover voltage may deviate from that of naturally polluted insulators. Recognizing this issue, IEC 60815-4 recommends that corrections be applied to account for the differing dissolution behavior of various salts when selecting or designing insulators for DC systems.

The primary aim of this chapter is to propose a method for evaluating and correcting the composition of soluble salts in naturally accumulated contamination, thereby improving the equivalence between SDD (used in artificial tests) and ESDD (measured under natural conditions). To achieve this goal, the following steps were undertaken:

1. The dissolution behavior of salts during ESDD measurements and under saturated moisture conditions was studied.
2. A correction methodology was proposed to adjust the soluble salt composition of natural contamination.
3. A measurement device was developed to assess the dissolution rate of contaminants on insulator surfaces under field-like conditions.

This research is expected to provide a practical and accurate tool for selecting appropriate insulators in high-voltage transmission systems and offer technical guidance for determining SDD values in artificial pollution tests.

14.2 Analysis of Surface Conductivity for Soluble Salt Detection

As outlined in the Introduction, the composition of soluble salts on naturally contaminated insulators is complex and is influenced by the interaction between environmental factors and insulator surface characteristics. For analytical simplification, this study uses sodium chloride (NaCl) and calcium sulfate ($CaSO_4$) as representative compounds for highly soluble and slightly soluble salts, respectively. These two substances are commonly found in natural pollution and exhibit distinct dissolution behaviors, making them suitable for investigating the variability between artificial and natural contamination conditions.

14.2.1 Dissolution of Soluble Salts During ESDD Measurement

To evaluate the soluble salt content on an insulator surface, standardized procedures such as the swab technique and washing technique are recommended by IEC 60815 and IEEE Std. 4. Both methods aim to extract and dissolve the surface contaminants using distilled or deionized water, followed by conductivity-based calculation of ESDD.

In the swab technique, the insulator surface is wiped using absorbent cotton soaked in distilled water, and the collected material is then rinsed into a water sample container. In the washing technique, typically used for cap-and-pin insulators, distilled water is directly poured or sprayed onto the insulator to wash off the surface contaminants, which are collected and analyzed.

Once collected, the electrical conductivity of the resulting solution—containing fully dissolved salts—is measured and corrected to the standard reference temperature of 20 °C. The ESDD is then calculated using the measured conductivity, the volume of distilled water used, and the cleaned surface area of the insulator.

To ensure consistent field measurement, the volume of distilled water used is standardized: For the swab technique, typical volumes range from 100 to 300 mL, depending on the insulator's surface area. For the washing technique, volumes of 500 to 1,000 mL or more are commonly used. In China, the swab technique is more widely adopted. For example:

For a standard cap-and-pin insulator (U70B/146, as per IEC), 300 mL of distilled water is typically used. For other insulator types, the volume of

Table 14.1 Volumes of distilled water recommended in China for ESDD measurement of various insulator surface areas.

Surface area (cm^2)	Water volume (mL)
≤1,500	300
(1,500~2,000)	400
(2,000~2,500)	500
(2,500~3,000)	600

distilled water is adjusted based on surface area, following practical guidelines. Table 14.1 summarizes the recommended water volumes for various surface areas. As can be observed, the average water usage is approximately 0.2 mL/cm^2, ensuring sufficient dissolution of even slightly soluble salts such as $CaSO_4$ under ESDD testing conditions.

It is well established that NaCl has very high solubility in water. At 20 °C, approximately 35.6 g of NaCl can dissolve in 100 mL of deionized water. By contrast, the ESDD of naturally contaminated insulators rarely exceeds 0.5 mg/cm^2. Even for large surface areas, the total salt mass typically remains well below the solubility threshold of NaCl in standard ESDD measurement volumes. Therefore, it is reasonable to assume complete dissolution of NaCl during the ESDD measurement process.

In contrast, calcium sulfate ($CaSO_4$), particularly in its hydrated form $CaSO_4{\cdot}2H_2O$ (gypsum), exhibits much lower solubility. The dissolution behavior of $CaSO_4{\cdot}2H_2O$ was experimentally examined, and for a volume of 100 mL deionized water, it can only dissolve 0.255 g of $CaSO_4{\cdot}2H_2O$ at 20 °C and the corresponding electrical conductivity is 1,760 μS/cm.

The relationship between ESDD and electrical conductivity is as follows:

$$\mathrm{ESDD} = \frac{S_a \cdot V}{A} = \frac{(5.7 \times \mathrm{EC}_{20})^{1.03} \cdot V}{A} \tag{14.1}$$

where EC_{20} is the electrical conductivity at 20 °C in S/m, Sa is the salinity of the suspension in kg/m^3 or mg/cm^3, V is the volume of the suspension (cm^3), and A is the area of the cleaned surface (cm^2). When the water volume used for ESDD measurement is 0.2 mL/cm^2, the maximum value of ESDD contributed by $CaSO_4{\cdot}2H_2O$ is 0.20 mg/cm^2 as calculated by Eq. (14.1). If the soluble salt in the contamination layer is only $CaSO_4$, it cannot be dissolved completely when the ESDD is greater than 0.20 mg/cm^2 ($CaSO_4{\cdot}2H_2O$ mass greater than 0.51 mg/cm^2).

As ESDD of the naturally polluted insulators is generally less than 0.5 mg/cm^2, all the soluble salts are completely dissolved, and therefore, the dissolution of soluble salts during ESDD measurement is 100%.

14.2.2 Dissolution of Soluble Salts Under Saturated Moisture Condition

In real-world outdoor environments, the wetting of insulators occurs primarily through natural processes such as dew formation, fog, or light rain, which result in limited surface moisture. According to the literature, the saturated water absorption of a typical contamination layer ranges between 2 and 8 mg/cm^2, forming a thin water film typically only tens of micrometers thick. This limited water volume significantly influences the dissolution behavior of different salt components on the insulator surface.

At 20 °C, the solubility of NaCl in water is approximately 356 mg/mL, and the density of water is about 1.0 g/mL. Based on this, it can be concluded that NaCl will dissolve completely under saturated moisture conditions as long as its surface concentration does not exceed 0.7 mg/cm^2. This threshold is well above typical field-observed ESDD values, which rarely surpass 0.5 mg/cm^2. Therefore, in most practical cases, NaCl can be considered fully dissolved during natural wetting events.

In contrast, $CaSO_4 \cdot 2H_2O$ (gypsum) has very limited solubility in water—only 0.255 g per 100 mL at 20 °C. Under saturated surface moisture conditions, with water films amounting to just 2–8 mg/cm^2, the estimated dissolution of $CaSO_4 \cdot 2H_2O$ is only 5.1–20.4 μg/cm^2. This means that $CaSO_4$ will not fully dissolve unless its contribution to ESDD is less than 4.6 μg/cm^2.

In other words, when the mass of $CaSO_4 \cdot 2H_2O$ on the surface exceeds this threshold, a portion of it will remain undissolved under real atmospheric wetting. This discrepancy highlights the overestimation of conductive salts in standard ESDD measurements, where abundant water ensures full dissolution of both highly and slightly soluble salts. In contrast, during actual service conditions, only the highly soluble fractions contribute to the formation of a conductive water film and thus to pollution flashover risk.

This distinction underscores the necessity of correcting ESDD values when using them to simulate field-representative pollution in artificial contamination tests, particularly in dry inland or industrial environments where $CaSO_4$ is more prevalent.

Given that natural contamination on insulators typically consists of a mixture of highly soluble and slightly soluble salts, it is important to distinguish between the total soluble salt content measured in the laboratory and the portion that actually contributes to surface conductivity under real atmospheric wetting conditions.

To quantify this, the dissolution ratio of soluble salts, denoted as k_s can be defined as:

$$k_s = \frac{\text{ESDD}_{\text{sol}}}{\text{ESDD}} \tag{14.2}$$

where

ESDD_{sol} is the salt dissolved under saturated moisture condition,
ESDD is the salt dissolved during the ESDD measurement procedure.

By definition, ESDD_{sol} is the parameter that more accurately reflects the actual conductive salt load relevant to the pollution flashover performance of insulators. In contrast, the standard ESDD value includes both highly soluble salts (e.g., NaCl) and slightly soluble salts (e.g., $CaSO_4$), all of which are fully dissolved due to the excess water used during testing.

As a result, the measured ESDD is typically greater than ESDD_{sol}, and the dissolution ratio k_s is less than 1. This highlights the potential overestimation of pollution severity when using uncorrected ESDD values in artificial pollution tests, especially for environments where slightly soluble salts are dominant in the natural contamination layer.

In the following parts, the dissolution amounts of soluble salts during the wetting process are analyzed.

14.2.2.1 Highly Soluble Salts in Contamination

As mentioned above, the highly soluble salt NaCl in the contamination can be completely dissolved under the saturated moisture condition. Thus, the relationship between the surface conductivity of insulator and ESDD can be obtained by the following derivation:

The mass of NaCl m_{NaCl} (mg) accumulated on unit area of insulator surface can be given by:

$$m_{\text{NaCl}} = \text{ESDD} \times A \tag{14.3}$$

where A is the unit area (1 cm^2), assuming that the volume of solution on unit area (1 cm^2) of insulator surface is V_{sol} (cm^3), then under saturated moisture condition, the surface conductivity of insulator K_{20} can be expressed as:

$$K_{20} = f\left(\frac{\text{ESDD} \times A}{V_{\text{sol}}}\right) \times \frac{m_{\text{sol}}}{\rho_{\text{sol}}} \tag{14.4}$$

where m_{sol} is the mass of solution on unit area (g/cm^2) and ρ_{sol} are the density of the solution (g/cm^3) at 20 °C, respectively. Figure 14.1 displays the relationship between the surface conductivity and water absorption of contamination layer at various ESDD. In Figure 14.1, the water absorption of the contamination

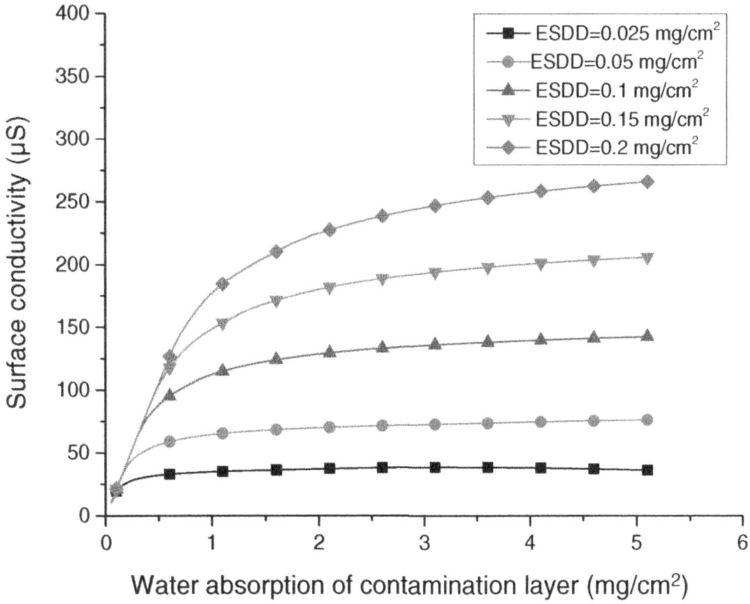

Figure 14.1 Relationship between surface conductivity and water absorption of contamination at various ESDD. *Source*: Cao et al. (2019)/with permission of IEEE.

layer surface is within the range of 2–5 mg/cm^2. The surface conductivity is not largely affected by the water absorption of contamination layer since NaCl is a strong electrolyte.

Figure 14.2 shows the relationship between ESDD and surface conductivity of the contamination layer under saturated moisture condition. It can be found that the surface conductivity of the contamination layer under saturated moisture condition increases with the increase of $ESDD_{sol}$. However, due to the interaction between ions in the contamination layer solution, the surface conductivity is not strictly linear with the concentration of NaCl solution. For the convenience of engineering applications, the relationship between surface conductivity and ESDD is can be exponentially fitted as

$$K = 1014.4 \times ESDD_{sol}^{0.8809} \tag{14.5}$$

14.2.2.2 Slightly Soluble Salts in Contamination

The effects of slightly soluble salts on the surface conductivity are analyzed in the same way as that for highly soluble salts. The only difference is that the solubility of $CaSO_4$ is much lower.

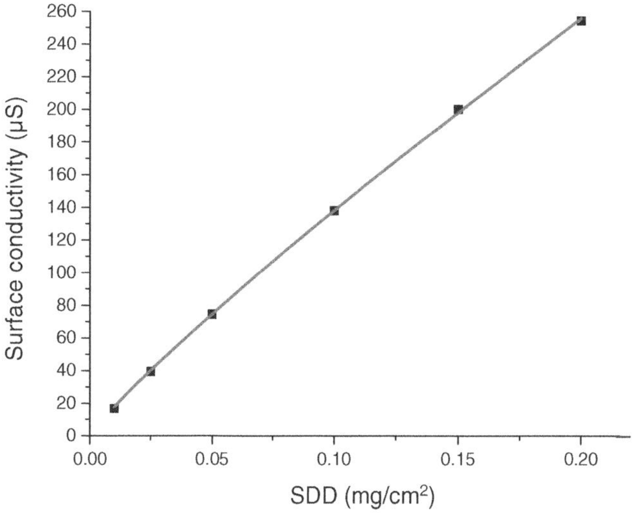

Figure 14.2 Relationship between ESDD and surface conductivity of the contamination layer under saturated moisture condition. *Source*: Cao et al. (2019)/with permission of IEEE.

When the concentration of $CaSO_4$ in the contamination solution is beyond a critical value, the electrical conductivity of the $CaSO_4$ solution reaches a saturation value. Then, the surface conductivity of the $CaSO_4$ contamination is:

$$K = EC_{max} \times h = EC_{max} \times \frac{m_{water}}{\rho} \tag{14.6}$$

where EC_{max} is the electrical conductivity of saturated $CaSO_4$ solution which is 1,760 μs/cm at 20 °C.

When the water absorption of contamination layer is 5 mg/cm^2, the relationship between the surface conductivity and ESDD is displayed in Figure 14.3. It can be seen that, for pure $CaSO_4$ contamination, the surface conductivity contributed by $CaSO_4$ is only 8.8 μS at 20 °C, when the ESDD of contamination is equal to or greater than 0.0046 mg/cm^2. A similar phenomenon occurs for other slightly soluble salts that partially dissolve in the saturated moisture condition, such as $Ca_3(PO_4)_2$.

14.2.2.3 Calculation of Dissolution Rates of Soluble Salts in Contamination Under Saturated Moisture

From the above analysis, it is found that the surface conductivity of insulators under saturated moisture condition is actually a reflection of the dissolution amount of soluble salts in the contamination. The larger the dissolution amount of soluble salts in contamination after saturated wetting, the higher the surface conductivity.

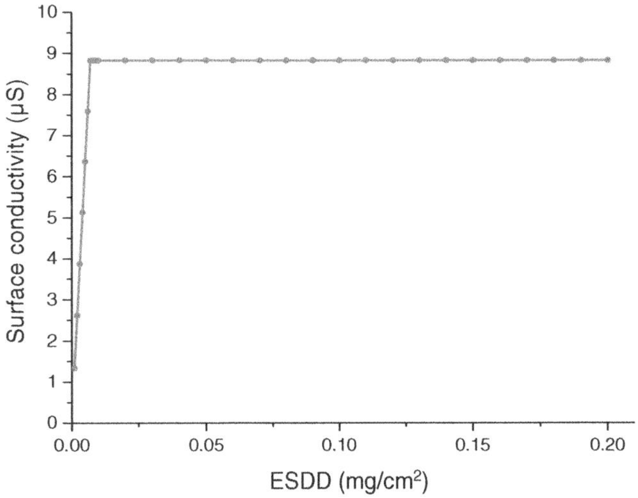

Figure 14.3 The relationship between the surface conductivity of insulator with calcium sulfate and ESDD (water absorption of contamination layer is 5 mg/cm^2). *Source*: Cao et al. (2019)/with permission of IEEE.

On the other hand, there is a relationship between the surface conductivity and the ESDD for using NaCl as highly soluble salt. Hence,

$$\mathrm{ESDD}_{\mathrm{sol}} = \frac{K^{1.14}}{2586} \tag{14.7}$$

Substituting Eq. (14.7) into Eq. (14.2),

$$k_s = \frac{K^{1.14}}{2586\ \mathrm{ESDD}} \tag{14.8}$$

Therefore, the parameter k_s, which takes into account both slightly and highly soluble salts, can represent the dissolution rate of soluble salts in the contamination under saturated moisture condition.

14.3 Correction Method for Soluble Salt Components and Equipment

14.3.1 Analytical Methods

Based on the preceding analysis, it is evident that a practical evaluation of the soluble salt composition on insulator surfaces can be achieved by jointly measuring: the ESDD, and the surface conductivity under saturated moisture conditions. This dual-parameter approach enables estimation of the dissolution behavior of salts under realistic service conditions, helping to distinguish between highly and slightly soluble components.

However, due to the nonuniform distribution of natural contamination on insulator surfaces, and the fact that surface conductivity measured between the steel cap and pin may not accurately reflect the true conductivity of the contamination layer, a localized measurement method is necessary.

To address this, a simple measurement device is proposed in this study to measure: **Local surface conductivity** under saturated moisture conditions, and **Local ESDD** within the same area. This localized approach improves the accuracy of contamination characterization by focusing on specific, representative sections of the insulator surface. The step-by-step procedure for evaluating soluble salt parameters is illustrated in Figure 14.4 and described as follows:

1. Selection of Measurement Area: A specific region on the insulator surface is selected for analysis.
2. Surface Conductivity Measurement: The selected area is sprayed with deionized water at a volume of 5 mg/cm^2, simulating the saturated moisture condition. Surface conductivity is then measured using contact electrodes placed over the wetted region. Once the surface conductivity stabilizes (after a short absorption and ion diffusion period), the value is recorded as the local surface conductivity under saturated moisture.
3. ESDD Measurement: An additional 195 mg/cm^2 of deionized water is sprayed on the same area (for a total of 200 mg/cm^2), ensuring full dissolution of both

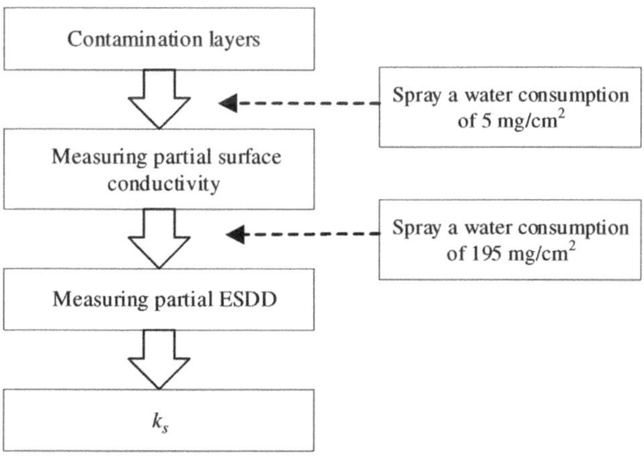

Figure 14.4 The measurement and analysis procedures to evaluate the contamination components. *Source*: Cao et al. (2019)/with permission of IEEE.

highly and slightly soluble salts. The water and dissolved salts are collected for local ESDD measurement.

4. Evaluation of Soluble Salt Composition: Using the relationship between the measured local surface conductivity (representing real atmospheric wetting) and local ESDD (representing total soluble content), the dissolution ratio k_s and relative contributions of highly vs. slightly soluble salts can be assessed.

It should also be noted that, during the ESDD and surface conductivity measurements, the conductivity of the solution was converted to the conductivity at 20 °C first by the formula in IEC 60815.1.

14.3.2 Measurement Probes

To facilitate accurate measurement of local surface conductivity and local ESDD under controlled wetting conditions, a custom-designed measurement device was developed. The equipment consists of four main modules, as illustrated in Figure 14.5.

The Water Injection Module is designed to deliver a precise and uniform application of deionized water onto a predefined surface area. Controlled dosing enables the application of specific quantities, such as 5 mg/cm^2 to reproduce saturated moisture conditions, followed by an additional 195 mg/cm^2 to ensure complete dissolution of surface salts during ESDD testing. Following wetting, measurement probes are positioned in contact with the surface to assess surface conductivity. The probes are constructed with carefully selected materials and geometrical spacing to minimize contact resistance variability, suppress parasitic effects, and maintain reliable electrical coupling, even on curved or textured

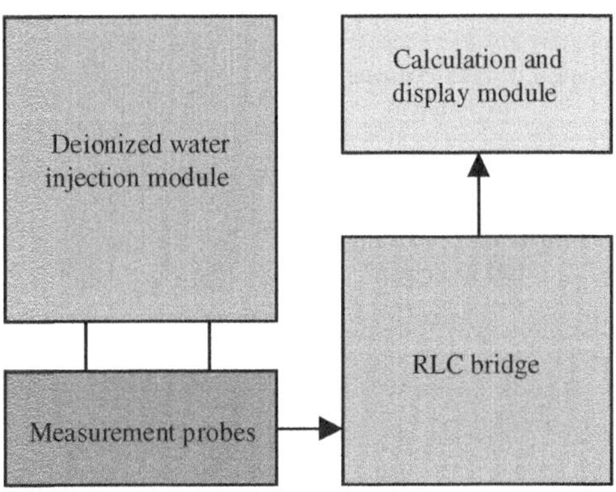

Figure 14.5 Schematic diagram of the measurement equipment.
Source: Cao et al. (2019)/with permission of IEEE.

insulator profiles. Electrical impedance between the probes is determined using an RLC (Resistance, Inductance, Capacitance) bridge. Given that surface conductivity is inversely proportional to surface resistance, the resistive component of the impedance is extracted and employed in the calculation of conductivity. The module demonstrates high sensitivity and precision under low-conductivity conditions, characteristic of the initial stages of wetting. Subsequently, the Surface Conductivity and ESDD Calculation Module processes the resistance data, transforming it into surface conductivity through established probe geometry and contact area parameters. After complete wetting and salt dissolution, the same module determines the local ESDD by integrating the total mass of dissolved salts, the applied water volume, and the effective cleaned surface area.

In this study, both local surface conductivity and local ESDD were measured using a specialized probe-based system. The surface conductivity measurement requires relatively low water consumption, using just enough water to ensure reliable electrical contact between the measurement electrodes and the surface contamination layer.

For this purpose, both plate and coaxial (concentric) electrode configurations have been shown to meet accuracy and stability requirements. These configurations are suitable for surface conductivity measurements due to their simplicity and direct contact interface with the thin water film.

However, when performing local ESDD measurements, where a larger volume of water (typically 200 mg/cm^2) is introduced to dissolve surface salts, measurement accuracy can be compromised by surface irregularities. This is especially evident near the edges or curved regions of the insulator, where small air bubbles may become trapped between the electrodes during liquid application.

In such cases, if a dual-plate or concentric circular electrode configuration is used, these trapped bubbles may accumulate between the electrodes, leading to inconsistent current paths and introducing measurement errors.

To address this issue, a two-cylinder electrode configuration is recommended, as illustrated in Figure 14.6. This design consists of two identical cylindrical copper electrodes placed vertically over the wetted surface.

When measuring the local ESDD, the relationship between the measured resistance and the conductivity of the contamination solution can be calculated numerically:

$$EC_\theta = K_f / R_\theta \tag{14.9}$$

where K_f is the shape factor of electrode. For the probe designed in this chapter, its radius was 5 mm, and the two symmetric copper rod electrodes have a radius of 0.5 mm, and a rod spacing of 7 mm. According to the calculation results, the shape factor of the probe used was 601.08.

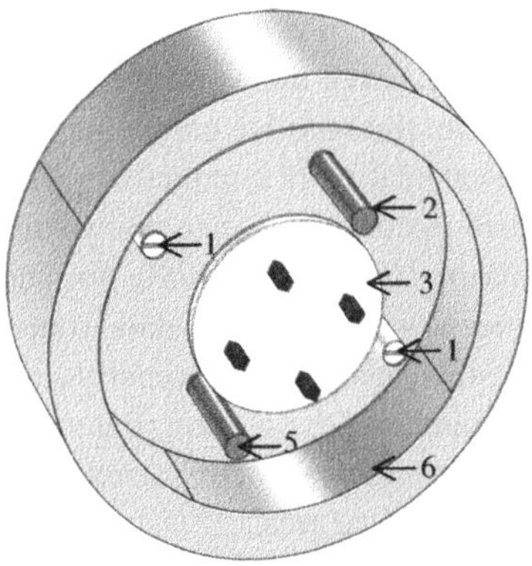

Figure 14.6 Design of measuring probes. 1,4: water inlet; 2,5: cylindrical electrodes; 3: dissolution accelerator; 6: inner wall. *Source*: Cao et al. (2019)/with permission of IEEE.

14.3.3 Water Consumption During Measurement

Affected by the insoluble material in the contamination, insulator material, and shed structure, the amount of water attached to the surface of different insulators under saturated moisture condition may vary slightly. When the water absorption of the contamination layer surface is within the range of 2–5 mg/cm^2, the surface conductivity is not largely affected by the water absorption of the contamination layer if the soluble salt is NaCl. However, for the slightly soluble salts such as $CaSO_4$, the conductivity will increase slightly with the increase of water consumption. To facilitate the repeatability and reproducibility of measurements, the water consumption for measuring local surface conductivity is specified to be 5 mg/cm^2, and the water consumption for measuring local ESDD to be 200 mg/cm^2 as summarized from Table 14.1.

14.3.4 $CaSO_4$ Dissolution Accelerator

During the ESDD measurement, it takes a long time to obtain stable values because of the slow dissolution of slightly soluble salts, which is unfavorable for field application. To accelerate the dissolution of slightly soluble salts such as $CaSO_4$ during the ESDD measurement, an insulated stirrer probe was added inside the electrode chamber. In order to avoid splashing of contamination layer solution, the probe speed was set at 20 r/min.

14.4 Experimental Validation of Correction Methods

To validate the proposed measurement approach and quantify the dissolution behavior of soluble salts under saturated moisture conditions, a series of controlled experiments were conducted using artificially contaminated glass specimens.

A pollution layer was uniformly applied to glass specimens with an area of 100 cm^2 using the solid layer method. The contamination mixture consisted of soluble salts—sodium chloride (NaCl) and calcium sulfate ($CaSO_4$)—and insoluble kaolin as a filler. The nominal salt deposit density was fixed at 1.0 mg/cm^2 across all specimens. For each experiment, the proportions of NaCl and $CaSO_4$ were varied, and the $ESDD_{sol}$ (salt dissolved under saturated moisture conditions) was determined based on the assumption that electrical conductivity is solely a function of ion concentration, given equal water volume. The precise mass of each salt used in the contamination mixture is summarized in Table 14.2.

A known quantity of contamination was weighed and thoroughly mixed with a measured volume of distilled water to form a suspension. This mixture was then evenly applied to the specimen surface using a fine brush to ensure uniform coverage. After air drying, the following measurements were conducted using the designed device described earlier:

Local surface conductivity under 5 mg/cm^2 water application to simulate saturated moisture, and local ESDD using an additional 195 mg/cm^2 of water to fully dissolve the salts. From these measurements, the dissolution ratio k_s was calculated for each case. The results are presented in Table 14.3, and it could be concluded that, the dissolution ratio varied with the contamination condition and reflected the degree of salt dissolution under saturated moisture.

Increasing the proportion of $CaSO_4$ in the contamination mixture results in a decrease in the dissolution ratio k_s. This observation aligns with the known low solubility of $CaSO_4$, which contributes less to surface conductivity when only minimal moisture is present. As a result, even if $CaSO_4$ constitutes a significant fraction of the total salt content, its limited dissolution under saturated moisture conditions causes a lower measured conductivity, thereby reducing k_s.

The dissolution ratio also varies with the total contamination level, represented by ESDD, even when the NaCl and $CaSO_4$ proportions remain constant. Specifically, at lower pollution levels (lower ESDD), the dissolution ratio k_s tends to be higher. This trend can be attributed to the higher water-to-salt ratio present in thin contamination layers. Under such conditions, even slightly soluble salts like $CaSO_4$ are more likely to dissolve fully, increasing the effective conductivity and the corresponding k_s value.

The relationship between ESDD and the surface conductivity are shown in Figure 14.7. It can be found that when the proportion of $CaSO_4$ is less than 20%, the contaminations are almost completely dissolved. With the increase of $CaSO_4$ proportion in the soluble salt contamination, the surface conductivity decreases at the same ESDD.

Table 14.2 The soluble salt components in the experiments.

ESDD (mg/cm^2)	The mass of NaCl and $CaSO_4$ measured in the experimental group (mg)											
	100% NaCl		80% NaCl		60% NaCl		40% NaCl		20% NaCl		0% NaCl	
	$CaSO_4$	NaCl	$CaSO_4$	NaCl	$CaSO_4$	NaCl	$CaSO_4$	NaCl	$CaSO_4$	NaCl	$CaSO_4$	NaCl
0.025	0	2.5	0.56	2	1.23	1.5	1.94	1	2.68	0.5	3.46	0
0.05	0	5	1.20	4	2.62	3	4.15	2	5.75	1	7.40	0
0.1	0	10	2.57	8	5.62	6	8.89	4	12.32	2	15.85	0
0.15	0	15	4.00	12	8.78	9	13.89	6	19.23	3	24.75	0
0.2	0	20	5.49	16	12.04	12	19.05	8	26.38	4	33.95	0

Source: Cao et al. (2019)/with permission of IEEE.

Table 14.3 The dissolution ratio of contamination (k_s).

ESDD (mg/cm^2)	The proportion of $CaSO_4$					
	0%	20%	40%	60%	80%	100%
0.025	1.00	1.00	0.82	0.68	0.47	0.24
0.05	1.00	1.00	0.75	0.57	0.35	0.12
0.1	1.00	0.98	0.81	0.56	0.33	0.07
0.15	1.00	0.93	0.77	0.57	0.31	0.05
0.2	1.00	0.93	0.79	0.58	0.31	0.04

Source: Cao et al. (2019)/with permission of IEEE.

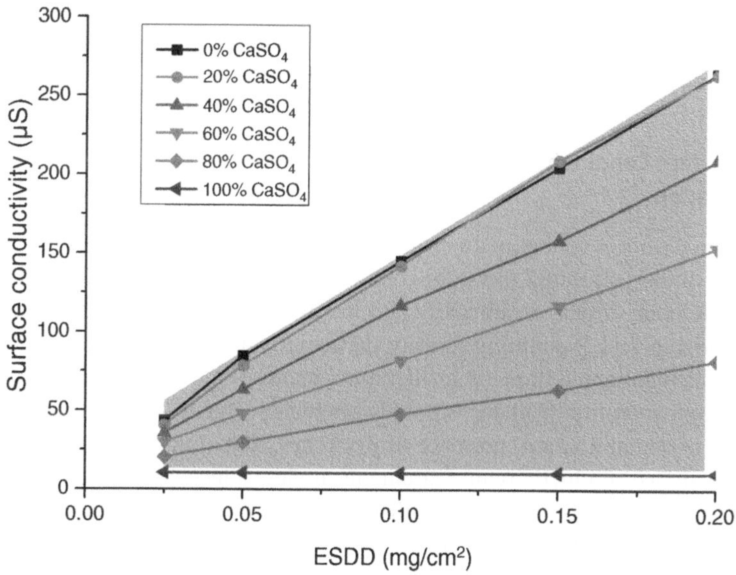

Figure 14.7 The relationship between ESDD and the surface conductivity. *Source*: Cao et al. (2019)/with permission of IEEE.

14.5 Engineering Implications

14.5.1 Surface Conductivity-based Soluble Salt Component Analysis

For naturally contaminated insulators, the surface conductivity under saturated moisture conditions consistently falls within a predictable range—represented by the shaded region in Figure 14.6. The upper boundary of this region corresponds to scenarios where all soluble salts are fully dissolved, as in the standard ESDD measurement procedure, which uses a substantially higher volume of water compared to natural wetting conditions (e.g., fog or dew).

Because the dissolution ratio of salts is higher during ESDD measurements than under saturated moisture, the measured surface conductivity in real service conditions will always be equal to or lower than that under full dissolution. This establishes a reliable envelope for expected values. Therefore, in cases where the soluble salt composition of surface contamination is unknown—such as insulators installed near uncharacterized pollution sources—an effective approximation of the proportion of slightly soluble salts can be made by measuring local ESDD and local surface conductivity, then interpreting these values using the established conductivity-ESDD relationship shown in Figure 14.7.

This method provides a practical tool for evaluating pollution characteristics in the field and can serve as a useful guide in the design and selection of outdoor insulation for transmission systems exposed to site-specific contamination environments.

14.5.2 Surface Conductivity-based Contamination Severity Characterization

The pollution flashover voltage of an insulator is strongly influenced by its surface conductivity under saturated moisture conditions. Recognizing this, IEC standards include both surface conductivity and leakage current as key parameters for characterizing Type B pollution severity, particularly in coastal environments.

However, as highlighted in previous literature, measuring overall surface conductivity across an entire insulator is not always reliable due to the nonuniform distribution of contamination, presence of dry zones, and structural variations. While local surface conductivity measurements are less affected by these issues, the process can be time-consuming, requiring multiple measurement points per insulator, making it less feasible for large-scale or routine field applications.

Despite these limitations, the local surface conductivity-based method proposed in this chapter—used to infer soluble salt composition—holds practical potential for industrial applications. Notably:

Frequent full ion analysis is not required in most engineering applications. Instead, a single-point local surface conductivity measurement can offer sufficient insight into contamination characteristics. Compared to advanced chemical methods such as ion chromatography, the proposed approach is simpler, faster, more cost-effective, and more directly related to flashover performance. Given these advantages, the method offers a promising alternative for evaluating contamination characteristics in a manner that is both practically relevant and technically grounded. Further development and field validation are recommended to support its broader adoption in the power industry.

14.6 Summary

In this chapter, the differences in the dissolution amounts of highly and slightly soluble salts under saturated wetting of insulators and during ESDD measurement were analyzed. In addition, a method for correcting the soluble salt components of natural contamination was put forward, and the corresponding measurement equipment was developed.

It was found that, for artificial contaminations dominated by NaCl, the contaminations are completely dissolved under saturated moisture condition. In this situation, the surface conductivity and the ESDD have a clear corresponding relationship. However, due to the low solubility of $CaSO_4$, the maximum contribution of $CaSO_4$ to the measured ESDD value is 0.20 mg/cm^2 when the water consumption for ESDD measurement is 0.2 mL/cm^2. When the water adsorption of contamination layer is 5 mg/cm^2 under saturated moisture condition, the maximum contribution of $CaSO_4$ to the surface conductivity is only 8.8 μS.

The parameter k_s is introduced and used to describe the dissolution rate of soluble salts in the natural contamination. The greater the k_s, the higher the dissolution rate of soluble salts on the insulator surface under saturated moisture condition.

By measuring the local surface conductivity and ESDD, a general method to evaluate the components of soluble salt in the natural contamination is proposed. It can provide a reference for the selection and design of transmission line insulators.

Bibliography

Cao, B., Wang, L., and Yin, F. Measurement of saturated water absorption of the contamination layer deposited on insulator surface, IEEE Sens. J., 2019, 19 (22), 10804–10811. https://doi.org/10.1109/JSEN.2019.2932192.

Chisholm, W.A. Adapting satellite measurements of air pollution to select insulators, INMR World Congress, Vancouver, BC, Canada, 2013, pp. 208–244.

Chisholm, W.A. Selecting Insulators Using Satellite Measurements of Air Pollution. Insulators of INMR, June, 2016. [Online]. Available: http://www.inmr.com/selecting-insulators-using-satellite-measurements-air-pollution/.

Chisholm, W.A., Buchan, P.G., and Jarv, T. Accurate measurement of low insulator contamination levels, IEEE Trans. Power Del., 1994, 9 (3), 1552–1557.

Chrzan, K.L., Stec, C., and Farzaneh, M. Probes for spot measurement of surface conductivity on polluted insulators, IET Sci. Meas. Technol., 2007, 1, 295–300.

Chrzan, K.L., Vosloo, W.L., and Holtzhausen, J.P. Leakage current on porcelain and silicone insulators under sea or light industrial pollution, IEEE Trans. Power Del., 2011, 26 (3), 2051–2052.

Eish, T.D., Abed, Y.D. and El-Behairy, H.A. A comparison of contamination types for organic insulating materials, IEEE Proceedings of 1st ICPADM, Xi'an, 1985, pp. 350–353.

Farzaneh, M. and Chisholm, W. Insulators for Icing and Polluted Environments, Wiley, 2009.

IEC 60507-2013, Artificial Pollution Tests on High Voltage Insulators to Be Used on AC systems, 2013.

IEC TS 60815-1, Selection and Dimensioning of High-voltage Insulators Intended for Use in Polluted Conditions - Part 1: Definitions, Information and General Principles, 2008.

IEC TS 61245, Artificial Pollution Tests on High-voltage Ceramic and Glass Insulators to Be Used on D.C. Systems, 2015.

IEC/TS 60815-4, Selection and Dimensioning of High-voltage Insulators Intended for Use in Polluted Conditions-Part 4: Insulators for D.C. Systems, 2016.

Jiang, X., Zhao, S., Xie, Y. et al. Study on fog flashover performance and fog-water conductivity correction coefficient for polluted insulators, IET Gener. Transm. Distrib., 2013, 7 (2), 145–153.

Jin, L., Tian, Z., Ai, J. et al. Condition evaluation of the contaminated insulators by visible light images assisted with infrared information, IEEE Trans. Instrum. Meas., 2018, 67 (6), 1349–1358.

Liljestrand, H.M. and Morgan, J. Error analysis applied to indirect methods for precipitation acidity, Tellus, 1973, 31, 421–431.

Richards, C.S., Benner, C.L., Butler-Purry, K.L. et al. Electrical behavior of contaminated distribution insulators exposed to natural wetting, IEEE Trans. Power Del., 2003, 18 (2), 551–558.

Takasu, K., Shindo, T., and Arai, N. Natural contamination test of insulators with DC voltage energization at inland areas, IEEE Trans. Power Del., 1988, 3 (4), 1847–1853.

Taniguchi, Y., Arai, N., and Imano, Y. Natural contamination test of insulators at Noto testing station near Japan sea, IEEE Trans. Power App. Syst., 1979, 98 (1), 239–245.

Wardman, J., Wilson, T., and Bodger, P. Volcanic ash contamination: limitations of the standard ESDD method for classifying pollution severity, IEEE Trans. Dielectr. Electric. Insul., 2013, 20 (2), 414–420.

Yang, Z., Jiang, X., Zhang, Z. et al. Study on the influence rules of soluble contaminants on flashover voltage of disc suspension insulators, IEEE Trans. Dielectr. Electric. Insul., 2016, 23 (6), 3523–3530.

Yuan, Q., Li, W., Zhou, S. et al. Integrated evaluation of aerosols during haze–fog episodes at on regional background in North China plain, Atmos. Res., 2015, 156, 102–110.

Zhang, Z., Jiang, X., Huang, H. et al. Study on the wetting process and its influencing factors of pollution deposited on different insulators based on leakage current, IEEE Trans. Power Del., 2013, 28 (2), 678–685.

Zhang, Z., Zhang, D., Jiang, X. et al. Study on natural contamination performance of typical types of insulators, IEEE Trans. Dielectr. Electr. Insul., 2014, 21 (4), 1901–1909.

Zhou, J., Xing, Z., and Deng, J. Characterizing and sourcing ambient PM2.5 over key emission regions in China I: water-soluble ions and carbonaceous fractions, Atmos. Environ., 2016, 135, 20–30.

Index

Printed and bound by CPI Group (UK) Ltd, Croydon, CR0 4YY
24/08/2026
14947829-0002